THE CHICAGO GUIDE TO YOUR CAREER IN SCIENCE

THE CHICAGO GUIDE

TO YOUR CAREER IN

science

A TOOLKIT FOR STUDENTS AND POSTDOCS

VICTOR A. BLOOMFIELD AND

ESAM E. EL-FAKAHANY

The
University of
Chicago Press
Chicago and
London

VICTOR BLOOMFIELD is professor in the department of biochemistry, molecular biology, and biophysics and a former dean of the Graduate School at the University of Minnesota. **ESAM EL-FAKAHANY** is professor of psychiatry, pharmacology, and neuroscience at the University of Minnesota Medical School. He is also former associate dean of the Graduate School. Together, El-Fakahany and Bloomfield established and directed the University of Minnesota's first office for postdoctoral affairs.

The University of Chicago Press, Chicago 60637

The University of Chicago Press, Ltd., London

© 2008 by The University of Chicago

All rights reserved. Published 2008

Printed in the United States of America

17 16 15 14 13 12 11 10 09 08   1 2 3 4 5

ISBN-13: 978-0-226-06063-7 (cloth)

ISBN-13: 978-0-226-06064-4 (paper)

ISBN-10: 0-226-06063-2 (cloth)

ISBN-10: 0-226-06064-0 (paper)

Library of Congess Cataloging-in-Publication Data

Bloomfield, Victor A.

　The Chicago guide to your career in science : a toolkit for students and postdocs / Victor A. Bloomfield and Esam E. El-Fakahany.

　　p.　cm.

　Includes bibliographical references and index.

　ISBN-13: 978-0-226-06063-7 (cloth : alk. paper)

　ISBN-10: 0-226-06063-2 (cloth : alk. paper)

　ISBN-13: 978-0-226-06064-4 (pbk. : alk. paper)

　ISBN-10: 0-226-06064-0 (pbk. : alk. paper)

　1. Science—Vocational guidance.　2. Research—Vocational guidance.　I. El-Fakahany, Esam E. II. Title.

Q147.B56 2008

502.3—dc22

2007038571

CONTENTS

PREFACE

Beginning scientific research as a graduate student or postdoctoral can be exciting, enriching, and the start of a rewarding career. But the world of research has become increasingly complex and competitive, and many students and postdocs in the biological and physical sciences make halting progress, don't find the jobs they want, and become disillusioned with research. *The Chicago Guide to Your Career in Science: A Toolkit for Students and Postdocs* addresses the problems of these beginning scientists in three ways. It explains

- how to navigate the system of the modern research university,
- how research time can be made more productive and motivation maintained, and
- how to develop communication skills and professional contacts.

There are many books that address graduate school, research, organization, motivation, communication, or careers, but none that covers all of these topics, particularly with a focus on beginning a career in scientific research. We have tried to integrate the practical advice and wisdom of these sources, augmented with our own extensive experience as scientists, mentors, and administrators concerned with graduate and postdoctoral education. We hope that students will find this to be a valuable guide at each stage of their training, and as they encounter the many activities that make up a scientific career.

In the twenty-first century, material is as likely to be published on the Web as in books and articles, but its permanence is less to be taken for granted. We have ascertained that all Web addresses included in this book were operational as of October 2007.

ACKNOWLEDGMENTS

We would like to thank Catherine Cioffi (project manager, Merck & Co.) and the following people at the University of Minnesota for their critical review of a draft of this book: Gail Dubrow (vice provost and dean of the Graduate School), Ilene Alexander (program director, Center for Teaching and Learning), Launa Lynch (postdoctoral associate, Cancer Center), and Shana Watters (graduate student, Department of Computer Science). We are also grateful to several anonymous reviewers for very helpful comments and suggestions.

We thank Carolyn Chalmers (Office for Conflict Resolution) and Jan Morse (Student Dispute Resolution Center), both at the University of Minnesota, for advice about how to handle problems in the classroom.

We are grateful to our editors at the University of Chicago Press: Christie Henry, whose enthusiastic support of this project buoyed our spirits, and Joel Score, whose careful and insightful copyediting made the book much more readable than it would otherwise have been.

INTRODUCTION
THINKING ABOUT A
RESEARCH CAREER

Do you want to do scientific research? If so, this book will help you answer two big questions:

- How can you prepare for a career in research or for an alternative career that uses a background in research?
- How can you conduct and present your research most effectively?

These are complicated questions. But whether you're a beginning researcher or just thinking about going into research, this book will take you through them systematically, pointing out the choices and best practices that will make your career as successful and rewarding as possible.

Graduate students and postdoctorals (researchers with a PhD who are seeking further training before embarking on independent careers) learn many specific techniques and scientific concepts, but they don't always acquire the general tools of the trade that are crucial for success in research. A visit to any bookstore will reveal dozens of books concerned with finding the right job, boosting motivation, improving time management, organizing records, communicating effectively, and behaving ethically. These books, often shelved under the heading "self-help," are commonly shunned by students and their professors, but many contain ideas that we believe are valuable—and that are, in fact, evident in the working styles of the most successful researchers. Our intention is to translate these ideas into precepts directly applicable to the life of the scientist.

Although our primary audience is novice scientists—those still in training—we hope that more experienced researchers will also find much of interest here. Few of us, at any age, use all the tools available to us to maximize our effectiveness.

This book is well suited to self-study, but it could also serve as a textbook for courses on research skills or career paths. Indeed, we believe that universities should be offering such courses to their graduate students and postdocs. The book is divided into two main parts. Part I (chapters 2–9) discusses the various stages of a research career and the choices that must be made at each stage. Part II (chapters 10–25) addresses the many tasks involved in doing research and how they can be accomplished in an effective and responsible manner. In this introductory chapter, we provide a brief overview.

The rewards of a research career

Our modern society depends on research to cure diseases, abate pollution, increase supplies of food and energy, and provide insights into our relations with nature. The activities research entails can be mentally and emotionally engaging, among the most absorbing and fulfilling of human occupations. And researchers are generally adequately compensated, work with intellectually stimulating people, and enjoy high regard within society at large.

Some people have jobs in which they do nothing but conduct research—for example, working in a government or corporate lab. Some combine research with other activities: members of university faculties teach and administer programs, and physicians at university hospitals are also involved in clinical work. Still others work in nonresearch jobs that nonetheless utilize their scientific background and skills: as journalists, teachers in liberal arts or community colleges, and managers in enterprises ranging from biotech start-ups to large technical companies. All of these people can benefit from the knowledge, skills, and habits of mind acquired in pursuing a research career.

The challenges of a research career

HISTORICAL PERSPECTIVE

Educational and career opportunities in the sciences have changed over the years. A National Science Foundation report on U.S. doctorates in science and engineering in the twentieth century found rapid growth in both doctoral education and federal expenditure for research and development (R & D) in the middle years of the century (Thurgood, Golladay, and Hill 2006). During the 1970s, however, the economic impact of the Vietnam war led to severe cuts in R & D funding. This, along with a saturation of the academic labor market in most fields, caused a decline in the number of doctorates awarded. Since then, gains in R & D spending and a defense buildup have fostered a renewed increase in doctoral degrees in the sciences and engineering.

There have also been shifts in the types of employment sought by new PhDs. The percentage taking academic jobs declined from 67 percent in the early 1970s to about 50 percent by the end of that decade and stayed at that level for the remainder of the century. In contrast, the percentage of

PhDs who chose industry more than doubled, from 12 percent in the early 1970s to 27 percent in the late 1990s. The last three decades of the past century also saw a significant increase in the proportion of PhDs who continued their training as postdoctorals, both because postdoctoral experience increasingly became a requirement for good professional scientific jobs and because the number of qualified applicants for such jobs outstripped the demand.

The past decade has not been kind to the career aspirations of many young (and older) researchers. The poor economy has not only depressed industrial and government hiring of scientists but also—by lowering endowment yields and state tax revenues, and perhaps encouraging older faculty to delay retirement—reduced the number of faculty positions offered by private and public universities. Fortunately, the economy now appears to be improving, and with it employment prospects. Even so, difficulty in finding a suitable job remains a common frustration—often resulting in a prolonged period in relatively poorly paying postdoctoral positions. In this book we present a variety of ways in which you can maximize your chances of escaping that holding pattern and finding a rewarding job.

FINDING THE RIGHT JOB

Research can be deeply rewarding, but if you don't enjoy the repetition, uncertainty, frequent failures, and delayed gratification—or at least understand their inevitability in pursuit of your higher goals—it can also be frustratingly tedious. You may find, as many have before you, that although you love science, you become disenchanted with the routine of research. Fortunately, there are many other things that you can do with a research degree, but the sooner you decide that research is really not for you, the sooner you can start on another path. According to Monster.com, people these days change jobs an average of eight times by the age of thirty, and perhaps twenty times during their working careers. While too many jumps may cause employers to wonder whether a potential employee will stay long enough to justify their investment in training, some changes of direction are not unusual. For this reason, we'll tell you not just about research careers but about other careers for which your research training will provide the right start.

COPING WITH THE DEMANDS OF A LIFE IN RESEARCH

The main task in research is generally clear: complete the project, finish the dissertation, write the paper. But making time for thinking, reading, experimenting, observing, interviewing, analyzing data, consulting with colleagues, going to lectures, traveling to meetings, presenting talks and papers, and writing grant applications can leave the most dedicated young researchers baffled about how to fit it all in. Frequently they plow ahead, gathering data but neglecting other of these crucial activities. And there are invariably other demands on their time and energy, from outside the research project: class assignments, work as a teaching assistant, family responsibilities, recreation, and perhaps an outside job.

In this book we present strategies for planning and organizing your time so as to make your research efforts as efficient and fruitful as possible, while not neglecting the other important parts of life. This advice is aimed at beginning scientists at crucial stages of their careers: undergraduates considering graduate school; new graduate students choosing thesis advisors and research projects; graduate students already engaged in research, some of whom are also writing their theses or dissertations; and postdoctoral fellows, fresh from graduate training and eager to learn another area of research before embarking on independent careers as academic, industrial, or government scientists.

Each of these stages entails great changes in one's life. New graduate students face a transition from a relatively structured, learning-oriented undergraduate environment to one involving fewer courses and exams, greater and more varied responsibilities, less overt or explicit guidance about how to fulfill those responsibilities, and an emphasis on making original contributions to the field of study. Postdoctorals enter a period in which they are free to do nothing but research, which might seem to be an ideal situation. They may, however, receive little direction, and evaluations of their success will be based almost entirely on how productive they are in research. This can cause a lot of stress. In addition to these work-related changes, many young researchers have family obligations or relationships to which they should devote some attention, and they all need to maintain their physical fitness, to get a few hours of sleep each night—in short, to have a life!

Organization of this book

EARLY DECISIONS IN A RESEARCH CAREER

In preparing for your career, there are several points at which you should revisit the fundamental question of whether the life of a researcher is for you: first, when you decide whether to pursue a PhD, a necessary credential for most research positions; when you choose whether to continue your research training as a postdoctoral; and finally, when you complete a postdoc and must determine what sort of employment to pursue. At each stage you have other options. The decision calls for deep soul-searching and consultation with friends and mentors.

The chapters in part I address these and other choices that must be made as you embark on a research career, and will help you to prepare for alternative careers should the straight research path not seem the best choice.

Should you get a PhD? If you want the option of doing independent research, you will need to have a PhD. But getting a doctorate will probably require four to six years of your life, working hard for low pay. It's a big investment if you don't think it will serve as the basis for a satisfying career. In chapter 2, we analyze the options.

If you decide to get a PhD, your next important decisions are which graduate program to join and who within that program you will work with. You'll want to consider many factors: Is the program strong in the research specialty that interests you? Are faculty members in that specialty likely to be available to supervise your research? Do they have a good record of mentoring and placing their students? Will there be adequate financial support while you work on your degree? We discuss these and many other considerations in chapter 3.

Once you've enrolled in grad school, started your course work, and chosen a research advisor, you'll have to choose a project for your dissertation. One of the most striking things about research, which differentiates it from other work that may require equivalent intelligence and training, is that it involves exploring the unknown: one can discover something new, something no one else has previously realized or thought about. This can be a great thrill, but it also presents a challenge. Asking the right questions—guessing what will be a productive problem to investigate—is probably the skill that best differentiates the most successful researchers from the others. We discuss strategies for choosing a productive research project in chapter 4.

In graduate school you'll likely spend some time as a teaching assistant. Not only will this provide financial support, it will also give you valuable training should you decide on a career as a professor. In chapter 5 we examine strategies for successful teaching.

Should you take a postdoctoral appointment? As you approach the end of your PhD program, another big choice awaits: Are you sufficiently happy with what you've learned about life as a researcher that you want to continue down that path? If so, you should probably take a postdoctoral position, which will broaden your scientific horizons and make you a more viable candidate for a career in research a few years down the road. If not, then this is the time to look for a career that uses your training in other ways. If you can't immediately find the kind of job you want, or if personal relationships or other constraints preclude moving to take a job, you may proceed with a postdoc in any case. Selecting a postdoctoral program involves choices similar to those you made in relation to graduate school: which institution, which mentor, and which research project? Now, however, you'll have a much better idea of the scientific direction you wish to pursue and of leaders in the field whose prestige and connections may later help you to get a top-rank job. In addition to university positions, you can consider postdocs in government or industrial research laboratories, which may pay considerably more. Chapter 6 addresses these issues.

Should you choose an academic or nonacademic career? As you near the end of your postdoctoral stint, you'll need to decide whether the life of a researcher still enthralls you and, if so, whether to try for a position at a research university, at a smaller college that encourages research with undergraduates, or in industry or government. Chapters 7–9 will help you sort out these options and tell you how to go after the job you want.

DOING RESEARCH WELL

Doing research well is not easy. Like any other important activity, it requires training, discipline, attention to big principles and small details, and a willingness to stick with it through some difficult patches. If anything, it's probably harder now than it was when some older university faculty members were beginning their careers. Beginning scientists now are more likely to have significant family responsibilities, the literature is much more voluminous, competition for grants and recognition is more intense (there are a lot more researchers now, for better or worse), and research funding and regulations require more paperwork. In part II of this book, we offer guidance for coping with the many demands of professional and personal life.

RESPONSIBLE AND EFFECTIVE RESEARCH

Ethical conduct. Too often research ethics is glossed over as obvious or is accorded less importance than spending long hours at the lab bench, giving spellbinding talks, or publishing in prestigious journals. But unless you're satisfied that your work has been done as well as you can do it, and confident that you and your research collaborators have done nothing that you would be ashamed of if it were publicly disclosed, then you've not done justice to science, to the society that supports it, or to yourself. We discuss the responsible conduct of research in chapter 10.

The research notebook. If there is one thing that characterizes good scientific work, it is the keeping of a careful notebook. A notebook is not just a place to record raw data; it's a place to analyze the data, reflect on their meaning, and plan new experiments. It documents the details of an experiment, which allows you to repeat it, or to defend it if your results are challenged. It's the key repository for evidence of an original discovery or idea if you find something worth patenting. Bound paper notebooks have been the standard, but electronic notebooks are now sometimes used, and this use raises complex issues. Chapter 11 examines the many facets of keeping a good research notebook.

Working with others. The popular conception of a scientist is of a loner, someone who hides in the lab and avoids other people. In fact, however, scientific research is a highly social activity. As a researcher in training, you will interact with research advisors, colleagues in the lab, and outside collaborators. You will have to learn from and teach others, exchange criticism and ideas, work out what parts of the research program belong to each participant, and deal with authorship and intellectual property issues. It is also possible that you will have to blow the whistle on someone—possibly a friend, a collaborator, or even an advisor—whom you think may have committed research misconduct. In chapter 12 we discuss ways of working effectively with others.

MOTIVATION AND ORGANIZATION

Sustaining creativity. Research is a creative activity. But creativity isn't limited to discovering big stuff: "Eureka!" or $E = mc^2$. It's also figuring out how to write a useful piece of computer code, or selecting an appropriate DNA sequence for a genetic experiment. Being steadily creative in big and small ways is important to research, but it's not easy. There are many failures, progress occurs in fits and starts, and much academic research lacks clear goals. As a result, confidence and motivation flag and need to be refreshed. Chapters 13 and 14 offer useful approaches to fostering creativity, solving problems, and maintaining motivation.

Managing your time. Graduate students have to study for classes, fulfill their responsibilities as teaching assistants, get familiar and stay up to date with the literature in their fields, prepare, perform, and analyze experiments, write up their results for papers or conferences, attend lectures and research group meetings, learn about career opportunities, keep in touch with family and friends, get some exercise and some sleep . . . the list goes on and on. Postdocs may not be teaching or taking classes, but they still have to keep learning, and their research pressures and family obligations may be even greater. Learning to get organized and manage time effectively is critical to success in research, and to balancing research with the rest of life. We tackle these challenges in chapter 15.

Finding and keeping track of information. Researchers, like everyone else, are constantly bombarded by new facts, ideas, and concepts. The research literature is huge and doubles every few years, which makes finding and keeping track of pertinent information particularly difficult. Computerized databases and Internet searches help somewhat, but they also make us painfully aware of how much is out there that we cannot afford even to pay attention to, let alone learn and assimilate. Nonetheless, researchers must keep up with new information in their fields and neighboring ones. Not only does their own progress depend on making use of others' findings, but success in research is determined by priority. If you do something apparently new and exciting but it turns out that someone else did it first, then that person gets the credit and you may be accused of inattentiveness at best or intellectual piracy at worst. Chapter 16 describes effective ways to find, organize, and recall scientific information.

COMMUNICATION SKILLS

Most modern fields of research are so specialized, so far from the commonplace world that posed the initial question, that their results are comprehensible only to specialists. Since a successful career—hiring, promotion, tenure, funding of research grants—generally depends on being judged by at least some people who are not specialists, learning to communicate effectively with nonspecialists is an important skill. Even other specialists can be so busy, tired, and distracted as to miss the significance of a new piece of information unless it is presented clearly and forcefully. Communication of complex information and ideas requires special effort and at least some training, adding to the demands of an already busy research life. In chapter 17 we give an overview of the importance to the researcher of effective communication, then in the remaining chapters discuss specific aspects of presenting your work.

Attending scientific meetings. A scientific meeting offers a valuable opportunity to let others in your field know what you're working on and what you've done, and to get feedback, critiques, and ideas from them. It's also a chance to learn about others' work and about new techniques and equipment that may be useful in your research. Meetings are also great places to network, to meet the leaders in your field—people who might become your postdoc mentor or help you find a job. Chapter 18 tells you how to use your time at a scientific meeting most productively.

Preparing effective posters. In recent years, as scientific meetings have gotten larger and the number of simultaneous sessions has grown to the point that few can attend all of the talks relevant to their interests, the poster session has become a prime means by which researchers present their work to their colleagues. The virtue of a poster is that you can spend considerable time going over your work with scientists who are particularly interested, rather than giving a superficial précis to a broad audience in a fifteen-minute platform talk. However, attractively summarizing your work on a poster board—generally no larger than the top of a desk—in a form that will draw the attention of those among the milling throng who should know about your research, is not a trivial task. Chapter 19 tells you how to create effective poster presentations.

Talking about your research. If people are to know about your research, you need to talk and write about it. Talking should come first; it's less formal and allows exchange, correction, and refinement of ideas before the results are committed to print. Talking about your research is also a way of interacting with people who may offer you your next job, so it's important to learn to do it well. In chapter 20 we discuss the many formats in which you need to be able to summarize your research—from the thirty-second "elevator talk," through the fifteen-minute platform talk at a meeting, to the hour-long formal lecture—offering strategies for effective communication in each format.

Learning to write well. Your research isn't really finished until it's written up, reviewed, and published. The published literature of science is the ever-growing foundation of future progress and the means by which your own contributions to human knowledge will be recorded and acknowledged. How widely your work is recognized, understood, and used by others depends in large measure on how well it's presented in writing. Unfortunately, many bright young scientists who have no trouble planning and carrying out elaborate experiments, statistical analyses, and calculations find that writing up their results is a difficult and painful experience. In chapter 21 we show you how to make writing about your research easier and more productive: how to use writing as an aid to thinking, how to quickly write first drafts and then revise for a finished product, how to overcome writer's block, and how to work with coauthors.

Preparing graphs and tables. It's said that a picture is worth a thousand words, and nowhere is that more true than in science. A well-prepared graph, a thoughtful diagram of a proposed mechanism, or a clearly organized table can make your results and ideas much more accessible than pages of text presenting the same material. Preparing a good graph or table may, however, take as much time and effort as writing a thousand words. In chapter 22, we tell you how to design your graphs and tables so that they communicate as effectively as possible.

Writing your dissertation, journal articles, and grant proposals. Researchers spend a large part of their time on two kinds of writing: journal articles presenting their results to other scientists and grant proposals directed toward funding agencies and foundations. Graduate students must produce a third type of writing: the thesis or dissertation. Although each of these formats uses the basic skills of scientific writing and graph and table design, each has its own stylistic conventions and requirements, which we discuss in chapters 23–25.

In summary, embarking on a career in research poses some difficult choices and tests your ability to work creatively and efficiently. But nothing worthwhile is ever accomplished without overcoming challenges. We think that this book will provide you with the tools to meet those challenges and achieve success in research.

Take-home messages
- Research can be an exciting career with rewards that include the discovery of new knowledge, but it is not for everyone.
- A research career is demanding and competitive. To be among the best you must be prepared to make personal sacrifices.
- Success in research requires both good graduate training and personal and professional skills that might not be part of your formal graduate education.

References and resources

Goldsmith, John A., John Komlos, and Peggy S. Gold. 2001. *The Chicago Guide to Your Academic Career: A Portable Mentor for Scholars from Graduate School through Tenure.* Chicago: University of Chicago Press.

Thurgood, Lori, Mary J. Golladay, and Susan T. Hill. 2006. *U.S. Doctorates in the 20th Century.* Special Report. National Science Foundation Division of Sciences Resources Statistics. http://www.nsf.gov/statistics/nsf06319/

THE STAGES OF A
RESEARCH CAREER

2 · PURSUING GRADUATE EDUCATION

Why go to graduate school? Perhaps you see it as a stepping-stone to a better, or better-paying, job than you could get with a bachelor's degree alone. Perhaps a suitable job hasn't turned up after college and you figure grad school is the next best thing. Or perhaps you are hungry for more advanced knowledge.

These are all understandable motives. But the decision to pursue a graduate education requires a great deal more study and planning. You should take an inventory of your interests and career goals, and analyze your personal skills and competencies. Being a graduate student is different from being an undergrad, just as college was different from high school. In grad school you will need to be much more self-reliant; the emphasis will be less on course work and more on self-directed activity, mainly research.

The main reasons to enter a PhD program in graduate school are to learn to do research or to qualify to teach in university or college-level programs that require a PhD. (We won't discuss vocational master's programs, which have a different focus.) If you have already conducted research and enjoyed it, or if you think you would, then grad school is definitely the way to go.

Points to consider before applying

Graduate school is not a place for people who are still trying to find themselves. Before applying, you should be certain of your career goals. And you should be aware of the sacrifices you will have to make. Graduate education will consume several years of your life (usually one to three years for a master's degree, five to seven years for a doctorate). In science and engineering it is common to pursue postdoctoral training for an additional three to six years prior to getting a job. This delay in getting your first job may create financial difficulties, even if your graduate program does provide tuition benefits and a stipend. You should be prepared to accept a lower standard of living, as you will most likely not have the time for a job on the side.

Graduate school requires dedication, energy, and hard work, in both course work and dissertation research. The nature and design of your particular research project will determine your daily and weekly schedule, but grad students and postdocs typically work late and spend some weekends and holidays in the laboratory or long periods in the field. Your work schedule will get even busier after you graduate. If you choose an academic career, for example, you will need to teach, do research, keep up with rapidly advancing knowledge, and serve on committees. There will always be competition between your job duties and personal life, and you will need to find a balance that works for you.

Graduate education is a must for doing research and can open the door to a career of discovery and invention. But a PhD cannot guarantee a better job in your particular field. If your main motivation is to create more or better career opportunities, a professional school might better serve your ambitions. Talk to your college career counselors.

SELF-APPRAISAL

It used to be that students entered graduate school with one common expectation: to work under a renowned faculty member, complete a dissertation based on a strong research project, and become a leading faculty member at a premier research university. Their thesis advisors had similar expectations. Over the years, though, the research enterprise has expanded rapidly—producing, among other things, large numbers of highly qualified PhDs—while the number of faculty positions at major research universities has not. Not all doctoral students can go on to duplicate the careers of their star research professors. But this is not the end of the world.

Students have increasingly realized that there are attractive alternatives to faculty positions in major research universities. In fact, some feel they are better off teaching in smaller universities or colleges, working in industry or government, being part of a start-up company, or going into journalism, technical writing, or patent law. These jobs may promise less pressure to get research grants, fewer working hours, better pay, desirable teaching opportunities, or a deeper understanding of how the scientific discipline affects real people—not to mention more time for oneself and one's family.

Graduate schools and their faculties are also increasingly recognizing these alternatives as good options for their students and for society. While there is considerable variation between institutions and departments, many are paying more attention to preparing their students for alternate career paths.

Take some time, then, to assess your own interests. Your ideas about your strengths, needs, and career goals are almost certain to change with time, but the exercise is worth the effort. The Rackham Graduate School at the University of Michigan has put together the following list of questions that you might ask yourself (Rackham 2004):

- What are my objectives in entering graduate school?
- What type of training do I desire?
- What are my strengths?
- What skills do I need to develop?
- What kinds of research or creative projects do I want to work on?
- How much independent versus hand-in-hand work do I want to do?
- What type of career do I want to pursue?

WHO CAN HELP YOU DECIDE?

Consult with career counselors, family, and friends. More importantly, talk to current graduate students, professors, and directors of graduate programs to find out what success in your field of interest really requires. If you can, try the work out. Volunteering on a research project as an undergraduate, for example, can help you decide, based on real-life experience, if a research career is for you. It also puts you in direct contact with people who are doing the sort of work you're considering. They can answer many questions, the most important of which may be whether they would do it again if they could turn back the clock. If you plan to go to graduate school directly after college, you should start this consultation as early as possible, no later than the middle of your junior year.

WHEN SHOULD YOU GO TO GRADUATE SCHOOL?

Many students pursue graduate education directly after college. By doing so, they maintain their momentum as students. They are also less likely to have extensive personal and family obligations. However, not taking a break may result in school burnout or financial difficulties due to accumulated debt. And some students need time to decide if graduate education is really for them or to take additional preparatory courses and standardized tests such as the Graduate Record Exam (GRE).

If you instead work for a few years after college, you can develop a clearer perspective on your educational and career goals. If the job is related to your potential area of study, you may also gain a deeper understanding of your field of interest and possibly generate research ideas with practical applicability. Sometimes, however, it can be difficult to give up a job, especially one that pays well, to go back to school, particularly when the decision affects other people in your life. If you do decide to take a break after receiving your undergraduate degree, to earn money or gain perspective on career choices, you should not be concerned that the interlude will damage your chances of admission to graduate school.

A MASTER'S OR A PHD?

A master's degree in science qualifies you for many interesting careers, for example, as a technical writer, foundation administrator, or teacher at a two-year college. There are also many well-paying industrial research jobs that require only an MS, though your potential for promotion may be limited. You can obtain a master's degree with or without a thesis based on a research project. The thesis may add one or two years to your time in school, but it also adds prestige to the degree. Moreover, it creates valuable networking opportunities and enhances your knowledge of the field and experience in writing and critical thinking.

If your career and intellectual ambitions involve becoming an independent researcher, however, you should choose doctoral studies. Admission to a doctoral program in the United States generally does not require first obtaining a master's degree. (This is not the case in some other countries.) From now on, we'll assume that you've chosen to seek a PhD, since this is the research degree.

Choosing a field of research

Now that you've chosen to go to graduate school to learn to do research, the next question is: research on what? Perhaps a problem you encountered while taking a course, writing a term paper, doing an undergraduate research project, or reading in a general science journal has piqued your curiosity. Or you may be intrigued by a question with broad or deep ramifications: How can one understand learning and memory? Transfer genes to cure genetic diseases? Visualize and manipulate single molecules? Predict earthquakes?

How do you decide what type of graduate program to apply to while you're still an undergraduate? Sometimes it is pretty obvious: if you're interested in a project involving organic chemical synthesis, you'll want to apply to chemistry departments. But suppose you want to study the genetics of neural development—what sort of program would best serve that goal? A genetics department? A neuroscience department? A psychology department? The demarcation of research expertise within and between departments is often unclear. For instance, a faculty member in pharmacology might investigate bacterial DNA repair mechanisms, while one in mechanical engineering might study factors affecting cerebral blood flow. You will also find that the faculty in a typical department have diverse educational backgrounds. A faculty member in biochemistry might have a PhD in biochemistry, molecular biology, physiology, pharmacology, genetics, or bacteriology. She or he might even have a PhD in computer engineering and study mechanisms of substrate-enzyme interaction using computer simulation. The point is that you need to look farther than the name of a department or graduate program; the goal should be to find faculty research expertise that matches your specific research interests.

Where to go to grad school?

Which are the best graduate programs? Generally, they're the ones that have the best research faculty in your field. If you hope to do important work yourself, you should try to become an apprentice to one of these leaders. However, identifying that person or program may not be as easy as it would have been a few decades ago. There are many more excellent departments, many more prominent, high-achieving scientists, and many more subdisciplines in each scientific field. It no longer suffices to say, for example, that Columbia is supreme in geophysics, or Berkeley in physical chemistry, or Indiana in genetics. There are many more centers of excellence, which gives you many more choices but also makes those choices more complicated.

There will usually be more than one scientist whose work you find particularly interesting, and they may not all be at the same institution, so you should apply to each of their schools in order to give yourself some choice. You're also covering your bets: if your first choice doesn't admit you, you've got several good alternatives to fall back on. Applying to five or six graduate programs is reasonable and probably quite common.

In some disciplines, you are admitted to the department and may have a year or so to choose an advisor, meanwhile being supported by a training grant or teaching assistantship. In other disciplines, you will be admitted to work with a specific advisor from the very beginning; indeed, the decision about your admission may be up to that person. In the latter case, you should focus more on the particular mentor than on the general program.

In either case, it's rare these days that the best person in a field is alone in a department, with no colleagues doing related work of comparable quality. While you might focus on a department because of a particularly prominent scientist, you should look thoughtfully at the work of his or her colleagues. Identifying a nucleus of people doing work in which you are interested ensures that you will have choices even after you are admitted to their department. This can be especially important if you find, for example, that the person you most wanted to work with has a full research group and can't take on more students, or has a reputation locally (if not nationally) as being unpleasant to work with or neglectful of students because of other professional commitments. Sometimes you will find a younger faculty member who is more ready to mentor you, more energetic and full of ideas, less distracted by outside commitments. Overall, a department that has multiple faculty members working in an area of interest to you is likely to be well equipped for the research you hope to pursue, to have other students and postdocs engaged in related projects, to bring in a steady stream of relevant seminar speakers, and be a lively place where you can develop as a scientist in a variety of ways. This general milieu will provide at least as much of your education as your research advisor does. As P. B. Medawar (1979) has written, "Isolation is disagreeable and bad for graduate students. The need to avoid it is one of the best arguments for joining some intellectually bustling concern" (p. 14).

How do you decide who are the leading researchers in a field? Through active inquiry. Ask your undergraduate professors what they think. If you get a chance to go to a scientific meeting as an undergraduate, take the opportunity to explore the opinions of people from other universities. Some of the leading faculty may themselves be at the meeting. Go to their talks, and try to chat with them outside the lecture hall. It will help if you have some thoughtful questions prepared, about either the talk or the field in general.

It's also a good idea to read papers written by the professors you are most interested in. The first step is to do a literature search, which these days is most readily done online using Current Contents, Medline, Inspec, PsychInfo, Google Scholar, or other databases devoted to professional specialties. Searching for an author will turn up a list of papers, generally with abstracts. Reading the abstracts for the past several years' publications will give you a good idea of how productive a research group has been recently (which may not correlate perfectly with its long-term reputation), what it's focusing on, and whether new directions are being explored. You can then pick a few interesting abstracts from each research group you're thinking of and find the full papers in the library or online. You could also send an e-mail to the authors asking for reprints, which are likely to be sent as PDF files.

Another way to learn what a research group has published recently, and what it is currently working on, is to visit its Web site. Usually the Web site will contain a summary of current research and citations of recent papers. The citations may include abstracts, and there may even be links to the full text of the papers. If the most recent citations appear to be a couple of years out of date, that may be evidence either that the group has not been active lately or that the director of the laboratory doesn't put strong emphasis on keeping the Web site current. Either explanation is cause for concern and further investigation.

The quality and originality of a researcher's published work is worthy of particular consideration. Each discipline has well-publicized "impact factor" rankings for the journals in its field, based on the number of times a typical article is cited in the year following publication. A listing of impact factors for journals in various fields can be found at http://www.sciencegateway .org/rank/index.html. There are also a few prestigious journals (most notably, *Science* and *Nature*) that publish articles in a variety of scientific fields. Citation indexes document how many times a publication has been cited by others—a good measure of its impact on the field. Very few scientists can claim the distinction of having a "citation classic," a publication that has been cited four hundred times or more. (In some fields with fewer researchers, one hundred citations might qualify an article for classic status.)

Other factors could also influence your decision on where to go to grad school. You may have family obligations or other constraints that tie you to a particular geographic region. Or if you're pretty sure you want a career in industry, you might want to limit your search to schools located in areas of the country that have a high concentration of pertinent companies (e.g., electronics in Silicon Valley or Austin, Texas; biotechnology in San Diego or Boston). Large firms recruit at top universities all over the country, but being located nearby may increase your chances of landing an internship, doing collaborative research, or developing contacts that will lead to employment upon graduation. At the very least, you'll want to be sure to look at programs in which the faculty have private-sector contacts or consultantships.

STAYING PUT

If you're already at a top-flight research university with a vigorous program in an area that interests you, then it might seem simplest and most sensible, in choosing a graduate program, just to stay where you are. At least in the United States, however, staying put has been discouraged by the top departments, which can have their pick of students from anywhere in the country and overseas. The usual rationale is that as an undergraduate you've already absorbed much of what the place has to offer, and you'd be better off learning something new somewhere else. It is not clear whether this is a totally valid argument. In the elegantly ironic words of P. B. Medawar:

> Such a choice will have the advantage that the graduate student need not change his opinions, lodgings, or friends, but conventional wisdom frowns upon it and is greatly opposed to young graduates' continuing in the same department; lips are pursed, the evils of academic inbreeding piously rehearsed, and sentiments hardly more lofty or original than that "travel broadens the mind" are urged upon any graduate with an inclination to stay put. These abjurations should not be thought compelling. Inbreeding is often the way in which a great school of research is built up. If a graduate understands and is proud of the work going on in his department, he may do best to fall into step with people who know where they are going. (1979, 12–13)

Attitudes toward staying put may be changing, even in the United States, as competition for top graduate students becomes more intense and as the complexity of graduate students' personal lives makes it increasingly difficult for many of them to relocate. Our own feeling, however, is that unless there are compelling reasons to stay where you are, travel really is broadening. Staying in the same place throughout the period of your early professional development can be intellectually narrowing, as well as depriving you of personal experience in the wider world.

Finding information about graduate programs

Once you've identified some key faculty with whom you might like to work, the next step is to learn about the graduate program in which they teach. In some fields (e.g., biochemistry, chemistry, physics) you will generally apply and be admitted to a graduate program without commitment to or from a particular faculty member. In other fields (e.g., ecology, engineering), you will apply with the understanding that, if you are admitted, you will immediately join a particular faculty member's research group. Be sure that you find out which system pertains in the discipline and departments you are interested in. It is best to discuss this issue with the directors of the graduate programs you are applying for. In either case, you may find that the faculty member you would like to apprentice with participates in two or more graduate programs. In this case you should select the program that is richest in faculty in your research area of interest.

Confusion sometimes exists about the difference between university departments and graduate programs. Each university organizes itself differently, but one can generally say that a department is the unit within which faculty are hired and hold their tenure, courses are taught, and undergraduates pursue their majors. Departments are usually grouped into colleges. For example, the departments of mathematics, chemistry, English, and history might be in the College of Letters and Sciences.

Graduate programs, on the other hand, often report to the graduate school of the university. (Since departments pay the faculty's salaries and remain their tenure homes, the authority of the graduate school may be tenuous but is real nonetheless.) A graduate program may involve faculty from a single department (as is often the case in mathematics or physics) or from several departments and colleges. A graduate program in neuroscience, for example, may involve faculty from neurology, physiology, biochemistry, cell biology, genetics, computer science, and psychology, as well as from an actual department of neuroscience.

For simplicity's sake, we will use "graduate program" and "department" interchangeably in what follows. But in exploring places to do graduate work, you should keep the distinction in mind and be sure to consider interdisciplinary programs as well as single departments. Recent years have seen an increasing mixing of disciplines and technologies to answer important scientific questions. Establishing new graduate programs provides a way for universities to move rapidly into new fields without undertaking the arduous task of revamping departmental structures.

Most graduate programs have brochures and posters with tear-off cards that you can use to request more information and application forms. Most schools now also have Web sites describing their graduate programs, with links to the home pages of individual faculty. You can locate these on the

Internet by way of the university's home page or, perhaps even faster, by entering, for example, "minnesota university computer science" in a search engine.

Your choice can be better informed (if not necessarily made easier) by resources such as Peterson's, which annually publishes guides to graduate research in the various disciplines, listing most of the PhD-granting programs, the faculty, and their research interests. Some discipline-specific societies (e.g., the American Chemical Society) go beyond this and include lists of recent publications and contact information. Peterson's has a Web site that can be searched for subject area, desired degree, and location (http://petersons.com).

Contacting and courting a prospective advisor

Admission to most graduate programs is a rather bureaucratic process. An admissions committee reads stacks of applications and makes decisions within bounds established by other bodies. The institution's graduate school, for example, has a formal role that usually involves enforcing minimum standards, checking that paperwork is completed, and collecting application fees, while legal requirements concerning equal opportunity and affirmative action prescribe a process in which favoritism is minimized.

Nevertheless, a potential faculty mentor can have considerable influence on the admission decision. This is particularly true in fields where faculty members choose, and tell the admissions director, which of the many applicants they would like to have join their research groups. This even allows the occasional student with terrible grades or GRE scores to be admitted on the basis of undergraduate research, a recommendation from a trusted colleague, or just a hunch that they might be really good. In practice, of course, applicants with high GPAs and GRE scores are typically the ones accepted, as they will have a better shot at fellowships, which take the financial burden off the mentor's research grants.

Even in fields where you're admitted to a program rather than to a particular research group, it can't hurt if a potential research advisor can be persuaded to put in a word on your behalf. The most effective way to persuade him or her to do so is to do something worthy of scientific notice. If your undergraduate research is published or accepted for publication before admission decisions are made, you might send reprints or preprints to prospective mentors. Include a note explaining that you're applying to their program and that you hope to continue a similar line of work under their tutelage. If you're presenting a poster at a scientific meeting, attract their attention to your poster and say that you look forward to continuing the interaction if you're admitted to their program. There's no need to be blatant in asking these researchers to pull strings on your behalf. If they are impressed, they'll get the hint.

It will do you no good to send a letter or e-mail saying simply that you've been very impressed with your would-be advisor's work and that you hope to work under his or her direction, without specifically mentioning what has impressed you. Prominent faculty members get many such entreaties each year, many of which are generic (most often from foreign students who don't know the customs of U.S. universities). At best, your message will be referred, without comment, to the secretary of the graduate admissions committee.

On the other hand, an e-mail or letter that describes—with some detail and cogency—your research interests, your career goals, and why your interests would fit with those of the advisor might make an impression. Having sent such a message, don't pester the recipient for a response. There are, however, other, more suitable ways of following up: you might ask to meet at a conference, send a reprint or PDF of a paper you've recently had accepted for publication, or comment *thoughtfully* on a recent paper by the potential advisor.

Maximizing your chances

Once you've decided on the graduate programs that best suit your purposes, you have to apply for admission. The best graduate programs and research groups are highly competitive. They have lots of applicants and can choose the best. Unfortunately, it's not always clear who the best candidates are; high undergraduate grades and test scores do not always translate into creative performance in research. Therefore, graduate programs have learned to take a variety of factors into account.

UNDERGRADUATE RESEARCH

Perhaps the biggest single factor in having a graduate program review your application favorably is having a notable record of undergraduate research participation and performance. It's good if you've spent a year or more—one term is barely adequate—in a professor's research group working on a research project. It's even better if you've presented the work at a scientific meeting, either regional or national, and better still if the work has been published in a peer-reviewed professional journal. Such a record shows that you have an interest in the field and the ability to do research—and to balance its time demands with course work and other parts of your life. The experience will also have put you in close contact with a professor who can write you a good, nonsuperficial letter of recommendation.

If you're applying to graduate school in a field such as ecology where students typically develop their own research projects, it can be important to have done some independent research as an undergraduate. Your demonstrated ability to function independently as a scientist, even on a small project of your own devising, will make you more desirable as a graduate student.

In any case, research experience will teach you particular skills that you may be able to use later. And it will familiarize you with the frontiers of research in a particular area, giving you a good chance to find out what kind of science interests you. You do not, however, have to continue that line in your graduate research.

GRADES

Good grades (B or better) are important, but perfect grades are not. A straight-A student who has done no undergraduate research is generally less appealing than a student with an A-minus or B-plus average who has completed some research. You should, however, try to get mainly A's in the science courses pertinent to your intended specialty. It's also expected that a student from a "weaker" school will have a higher GPA than a student from a "stronger," more highly ranked institution, where standards are presumed to be higher.

GRADUATE RECORD EXAM

Most, but not all, graduate programs require that you take the GRE. Attention is generally focused on the GRE General Test (verbal, quantitative, and analytical writing); applicants to the best programs will generally have scores in at least the 70th to 80th percentile (though unfortunately verbal scores are often lower for science students). There is, however, almost no correlation between General Test scores and performance in graduate school beyond the first year. (There is a better correlation with the GRE Subject Tests, but ironically these are seldom required.)

Opinions differ on whether studying for the GRE General Test is useful, but it's prudent to take it seriously. A good place to start is the Web site of the Educational Testing Service, which devises and administers the exam.

There you can learn about the test, register to take it on paper or online, and get access to test preparation materials. The next step is to get one or two test preparation books, such as those from Barron's, Kaplan, or Princeton Review, listed at the end of this chapter. The About: Web site also has useful information on the GRE and other topics related to graduate study. These resources provide practice tests and suggest study approaches. Especially important are practice in vocabulary building and strategies for quantitative problem solving.

LETTERS OF RECOMMENDATION

Letters of recommendation from professors who know you well are very important. Getting to know a professor well enough that he or she can write such a letter on your behalf is one of the main reasons for doing undergraduate research. (There are, of course, other ways to get to know faculty members, such as being active in student organizations for which they serve as advisors or meeting your instructors during office hours or in other out-of-classroom situations.) Such a letter can attest in specific terms to your intelligence, motivation, energy, agreeable personality, unusual capabilities, and particular knowledge. Here's one (with names and a few other details changed) that one of us wrote for a former student:

> I am delighted to write in support of James Jenkins for admission to your biophysics graduate program. Jim has been working in my lab for the past three years. He started out as a physics major but decided that the physical aspects of biochemistry were more interesting and promising. He wanted to get some firsthand experience in a lab, and our advising office referred him to me. I agreed to take him on as a volunteer, telling him that he could help with preparing solutions and cleaning up, and learn basic techniques and ideas from the grad students and postdocs. He did this with great enthusiasm and dedication, spending a lot of time in the lab and quickly convincing everyone that he was bright and helpful.
>
> He soon was enlisted by one of my postdocs in a collaborative project with Prof. Marilyn Smith in Cell Biology. The project involves preparation, purification, and physical characterization of plasmids with inserted repeat sequences that affect segregation and protein binding. He has found striking evidence that the existence and positioning of these inserts have unexpected effects on the ability of multivalent cations to condense the plasmids. Jim has gotten involved in plasmid preparation, light scattering, gel electrophoresis, and electron microscopy at a level

that I would consider normal for a graduate student, not for an under-graduate. He's currently writing up some of this work for publication; it has already been presented, or is scheduled for presentation, at several poster sessions locally and nationally.

Jim spends a lot of time in the lab, but he also spends enough time on his course work to get nearly straight A's. In his senior year he has also taken on some extra duties: head TA in our nonmajors undergraduate bio-chemistry course, and co-organizer of a new undergraduate Biochemistry Club. He's one of those extraordinarily talented undergraduates we get in the College of Biological Sciences at Minnesota every couple of years, who seems to do everything well while hardly breaking a sweat. At the same time, he's extremely modest and unassuming, and is a real pleasure to have around. I think he has a very promising future as a scientist, and that he's perfect for a first-rank biophysics graduate program such as yours.

If you don't take the time to ensure your professor knows you, you'll get a far more perfunctory letter:

Mr. X was a student in my Chemistry 1000 class. He received an A, ranking in the top 10 percent of five hundred students. He appears to get along well with others.

Which letter do you think will make a greater impression on a graduate admissions committee with many more applicants than it can accept? Such letters are also necessary in applying to medical school or other competitive professional schools.

PERSONAL ESSAY

Most graduate school applications require you to write a personal essay—a couple of pages describing your background, scientific interests, and career aspirations. Those we've read are occasionally impressive but more often callow and inarticulate. It's hard to write an essay that's other than vague and naively idealistic unless you've accumulated some real-life experience. It is not impressive to mention that you would like to study the mechanisms of Parkinson's disease because you have an aunt who suffers from the dis-ease, or to study physics because you have been intrigued by how earth's gravitational force is generated since you were a child. Undergraduate re-search, summer work in a scientific company, and other actual experience will enable you to provide some concrete details about what you've done and how that has made you interested in pursuing a career in research in your chosen field. (Another argument for doing undergraduate research!) You might, in fact, draw on your interests and experience to present some ideas about potential thesis topics. You won't be making any commitment, but this will give the admissions committee a clearer idea of what you are interested in and how you might approach research problems.

ENGLISH LANGUAGE COMPETENCY

This point is addressed particularly to international students. To be admitted to graduate school in most U.S. universities, you'll have to achieve a minimum score on the Test of English as a Foreign Language (TOEFL) or similar exam. Take our word for it: the bare minimum is not enough. Your chances of admission will rise considerably if you score well above the minimum (so long as your recommendations, grades, and other test scores are also good). If you are admitted with only minimal English language competency, you will have a very difficult time for the first year or so. You will have a hard time understanding lectures and conversations, your professors and fellow students will have a hard time understanding you, and you may not qualify for a teaching assistantship, an important source of financial support.

English has become *the* language of science. It's worth your while to spend considerable time and effort mastering it if you hope to have a significant career in science. Nearly all the significant scientific literature is written in English, international conferences are held in English, and researchers from countries around the world have English as their only common language. While many foreign students learn to read and write English in high school, they rarely get adequate training in conversational English from native speakers. If you have an opportunity to get such training, either in school or from a private tutor, your investment in time, effort, and money will be amply repaid.

It goes without saying that native speakers of English should have a high level of competency in their own language. Nothing makes a smart person appear foolish more readily than incorrect or inept use of their native tongue. Unfortunately, these skills cannot be taken for granted, particularly among science and engineering students, who may be more comfortable using math than English to make their case. The educational systems of the United States and Great Britain are not as successful as they might be in instilling high standards of literacy in their students.

RECRUITING VISIT

It has become increasingly common for departments to arrange recruiting visits for attractive applicants, either paying all the costs or hoping that travel costs will be shared among several geographically clustered institutions that the applicant may be visiting. These may be individual visits at the applicant's convenience, but often a number of applicants are brought in together for "recruiting weekends." This makes hosting easier for the department and gives the applicants a better sense of community, introducing them not only to the faculty and current graduate students but also to likely classmates.

If you're invited to visit a graduate program, with the institution paying some or all of the expenses, the odds are that the admissions committee has already decided to make you an offer. At the very least, you're on the short list. The position may be yours to claim, but you could lose it if you come across as uninformed, uninterested, and uninquisitive. Use the visit effectively: speak with the faculty you're most interested in, ask questions, talk about your research experience, and in general show that you're intellectually alive.

Some practical questions

When choosing a graduate school, finding a good intellectual and psychological "fit" is important. However, there are also practical questions that should not be overlooked. The program's recruiting brochure and Web site will provide some of the answers, but the recruiting visit is a good opportunity to probe more deeply. Be sure to speak with current students, who can give you information based on their own experience.

Does the program offer adequate financial support? The amount of money you need for living expenses will depend significantly on local housing costs. It will also depend on whether you have to pay tuition and fees (including health insurance) out of your stipend, or whether those are covered by the graduate program. Will the support continue throughout your graduate career so long as you are making adequate progress toward a degree, or is there some fixed cutoff date? You shouldn't make your decision based on a difference of a few hundred or a thousand dollars between programs, but you should be confident that you won't be distracted by monetary worries or need to get another job to support yourself. There are many online tools that offer useful comparisons of what your dollar is worth in different cities, along with information on crime rates, taxes, quality of life, schools, and so on. (For example, see http://www.moving. com/Find_a_Place/Cityprofile/.)

Is departmental support or a training grant available? In some fields, students are admitted to work with a particular professor and immediately go on the payroll for his or her lab. In other fields, particularly the lab-based biological sciences, it is common for entering graduate students to do a number of research rotations (typically three) in different laboratories while finishing their course work. Take advantage of this opportunity to make sure that you will be doing research on a topic you really like, and that you and your potential research advisor get along.

How much teaching will be required? Some teaching is useful; it provides experience in organizing and leading a discussion section, in answering questions on your feet, and in learning material in an active way. But too much required teaching, particularly in the later years of your graduate program, will intrude on your research.

Are the faculty in your area well supported by research grants? If they do not have grants from federal agencies or foundations, it may mean that their research is not highly regarded by peer reviewers. It may also mean that they will lack the resources to support you as a research assistant, in which case you may have to spend extra time as a teaching assistant.

How many students are in the program and in the research groups you're most interested in? Too many may mean that you'll have trouble getting into your preferred group, or have difficulty getting access to research resources or to your research advisor. Too few may be insufficient to maintain the intellectual energy of the environment or to provide a network of professional and social support.

Does the department or program have adequate facilities and instrumentation? If not, you may not be able to do the state-of-the-art work that you should be aiming for.

Is there a steady stream of prominent seminar speakers from outside institutions? Outside speakers will help you keep in touch with the latest developments, including findings that have not yet been published but may affect your research.

How much course work is required in your graduate program? Some departments have extensive, rigid requirements that take a long time to complete, time that might be better spent on research or independent learning. Other departments may have almost no requirements, so students can get into the lab early. If carried to an extreme, this can lead to a neglect of formal learning of the fundamentals of your discipline, which can be a handicap as you become independent and try to build your scientific career. Different specialties will have different needs in this regard, and different students will respond better to more or less structured programs. The important thing is to be aware of the options, and to be conscious of the choices you are making.

How long does it take to get a PhD? The national average is around six years in the natural sciences, longer in the social sciences and humanities. Some students, and some projects, will take longer than others, but try to get a sense of how long previous students in the program have taken and when the current ones expect to finish.

What kinds of positions do graduates of the program typically get? Do they land good postdoctoral positions that lead to academic or industrial research jobs? Do they have trouble finding professionally suitable permanent employment and end up in long holding patterns? The best programs will usually have the best placement records, but it may also make a difference whether they offer training in how to teach (some schools, for example, take part in the Preparing Future Faculty program), hold sessions on what types of jobs are available for graduates in their discipline, and provide help to develop job searching and interviewing skills.

What other support services are available at the university? The quality of the health service, libraries, computer labs, parking and transportation, disability services, and recreational facilities can all have a big impact on your graduate experience.

What is it like to live here? Is decent housing available at reasonable cost? Are there agreeable places for socializing? Is there adequate public transportation, or will you need a car? Is the community friendly to students? Is the campus a pleasant place to spend time? Do other students seem as if they'd be pleasant to be around? You'll be here for several years; you might as well enjoy it.

In general, most highly ranked graduate programs in the sciences will score well on these items. They'll provide adequate financial support, be reasonably flexible, not require too much teaching, maintain good grant support and modern facilities, bring in numerous outside visitors, and have sensible course requirements. But different programs have different characteristics, and you should pay attention to these factors as you investigate and visit potential graduate schools, to be sure that their style would mesh comfortably with yours.

Take-home messages

- Again, graduate education is not for everyone; assess yourself and your career objectives before applying.
- Start preparing to apply to graduate school as early as possible. There is much homework to be done and many requirements to be fulfilled.
- Choose a field of research that intrigues you. This will make the hard work associated with research worthwhile.
- An institution with a solid reputation in your field of study will offer you more opportunities to select an appropriate advisor and dissertation committee.
- Select a graduate program that offers you both financial and intellectual support.
- Currently enrolled graduate students are a valuable source of information about a program's pros and cons.
- Job placement of past graduates is a helpful measure of quality of graduate education.

References and resources

About:Graduate School. http://gradschool.about.com/cs/aboutthegre/.

Educational Testing Service GRE Web site. http://www.ets.org/gre/.

Green, Anna L., and V. LeKita, eds. 2003. *Journey to the Ph.D.: How to Navigate the Process as African Americans*. Sterling, VA: Stylus Publishing.

Green, Sharon W., and Ira Wolf. *How to Prepare for the GRE, Graduate Record Exam* (book and CD-ROM). Hauppauge, NY: Barron's.

Johnson, W. Brad, and Jennifer M. Huwe. 2003. *Getting Mentored in Graduate School*. Washington, DC: American Psychological Association.

Kaplan Educational Centers. *GRE Exam: Premier Program* (book and CD-ROM). New York: Simon & Schuster.

Medawar, P. B. 1979. *Advice to a Young Scientist*. New York: Harper & Row.

Princeton Review. *Cracking the GRE* (book and DVD). New York: Random House.

Rackham Graduate School, University of Michigan. 2004. "How to Get the Mentoring You Want." http://www.rackham.umich.edu/StudentInfo/Publications/StudentMentoring/mentoring.pdf, p. 7.

3 ADVISORS AND MENTORS

Graduate research, especially in the sciences, has generally depended on a strong interaction between a student and a faculty member who guides the student's thesis research. In the old days, this faculty member was called the thesis advisor or research advisor. Now there is a tendency to call such people mentors and to recognize that mentoring is a key aspect of success in graduate school and in subsequent careers. Students are increasingly urged to be proactive in getting the guidance they need, and faculty are expected to be more explicit and conscientious in their mentoring.

It is useful to distinguish between a research advisor—whom we will define as the faculty member who supervises your thesis research—and a mentor. As noted in the excellent University of Michigan Rackham Graduate School Web site,

> a mentoring relationship is a close, individualized relationship that develops over time between a graduate student and a faculty member (or others) that includes both caring and guidance. Although there is a connection between mentors and advisors, not all mentors are advisors and not all advisors are mentors.

Mentors, as defined by Zelditch (1990), are

> advisors, people with career experience willing to share their knowledge; supporters, people who give emotional and moral encouragement; tutors, people who give specific feedback on one's performance; masters, in the sense of employers to whom one is apprenticed; sponsors, sources of information about, and aid in obtaining opportunities; models of identity, of the kind of person one should be to be an academic.

Your research advisor may not give you all of the mentoring support you want and need. That's OK, as long as your advisor treats you with respect and gives you the appropriate help in developing your research and professional skills. Rather than expecting your research advisor to give you everything, find others—other faculty members, senior graduate students, postdocs, and departmental staff—who can provide additional guidance and support. Having multiple mentors is all to the good.

How to identify potential research advisors

The first and most important mentor you should choose is the faculty member who will advise you in your thesis research. (If you've been admitted to a program to work with a particular faculty advisor, this choice will already have been made.) You've presumably chosen your graduate program because it has a strong faculty in your area of interest. You've read some of those professors' recent papers before making your choice, and you may have talked with several faculty members during a recruiting visit to the campus.

You need to recognize that the choice of research advisor will not be entirely yours: a research group may be full or have several students competing for one or two vacancies, there may not be funding to support you as a research assistant, or the person you'd like to work with may be going on sabbatical or have some other reason for not taking on new students. Nonetheless, this is the time to begin the detailed and careful investigation that will enable you to make the best possible choice. There will generally be an orientation period, either before classes begin or spread out over the first few weeks of the fall term, during which faculty members give brief, general talks about their research interests and invite questions and office visits from interested students. In this chapter, we'll give you a list of questions that you should ask the professors, their students, and knowledgeable others.

Different graduate programs have different timetables for choosing a research advisor. In some, you may make the choice when you are admitted to the program, in others a few months or a year may go by, and in others several years can pass before a thesis advisor is selected. Be sure to find out what the expected timetable is in your program at your institution, by reading the graduate program handbook in print or on the Web, and by talking to knowledgeable people. If you don't have to choose an advisor before classes start—because there are departmental funds to support first-year students or because you have a teaching assistantship—you may take classes from prospective advisors to get a sense of their intellectual style and how they relate to students. If one of them gives a seminar on his or her research in the department or university, be sure to attend. Social interactions with other students will also give you a chance to gather information. Ask more advanced grad students and postdocs, both those who work in the labs of potential advisors and those who are in related areas, what they know about the research groups that most interest you. In the biological sciences, it is common to have a series of research rotations, each lasting a few months, in different laboratories during the first year of graduate school. These will give you the opportunity to make sure you are getting into a research field that you really like, as well as the chance to find out how you and your potential research advisor get along.

This is also a good time to refresh your self-appraisal about what you want from graduate school (see chapter 2). Is it most important to you that your research lab has a stellar reputation, that it has strong industrial ties, or that your faculty advisor is especially supportive of your interests in college teaching or in nonacademic work? You may be wise to keep these alternative career interests to yourself for the time being, since not all faculty are happy with the idea that a student is interested in anything but research, but you should keep them in mind as you explore.

CHOOSE A FAMOUS RESEARCHER?

If you followed our advice in choosing a graduate school, you went to a place that had one, or preferably several, of the leading scientists in your area of interest. It is from among these people that you should try to choose a research mentor. As J. E. Oliver says in *The Incomplete Guide to the Art of Discovery*:

> Scientific research is mostly learned from the mentor. . . . The overall curriculum in graduate school is less critical. . . . What is most important in graduate school is to be closely associated with the leader of the field. Only through close association is it possible to learn how he or she thinks, plans, and operates in order to reach and maintain that position of leadership. . . . A student with only modest talents can be inspired and boosted to important achievement by association with the right leader. And, of course, in turn a leader will flourish through association with good students. (p. 81)

Oliver qualifies his advice, however, and urges caution:

> An occasional great scientist is eccentric to the point of avoiding any interaction with students. And some highly rated ones have already moved beyond their major discovery period to other activities. Some may not be properly rated. (p. 83)

Leaders in the field should be effective advisors and mentors for a number of reasons:

- They are likely to have deep insight into the outstanding problems of the field and thorough knowledge of useful approaches, as well as experience of what hasn't worked (an expert has been defined as one who has made all possible mistakes in an area!). This is important for the choice of problems, for pursuit of their solutions, and for general inspiration.
- They may have unique intellectual or personal styles. Being in close touch with an outstanding person, at an impressionable time in your life, can have lasting effects.
- They have broad contacts and can put you in touch with other leaders in the field, both now and in future job searches.
- The other students and postdocs in their labs will be of high quality. You will do much of your learning from them.
- The lab will be well equipped and well funded, allowing you to use cutting-edge techniques and not be unduly hampered by financial restrictions.

There are, however, some downsides to choosing a prominent scientist as advisor:

- Leading scientists will be very busy on the national and international scene, unless they protect their time unusually well. They may be away from campus often and have little time for their research students.
- Their research groups are often large, and they may have little time to devote to you and your problems until you come up with research results that catch their interest.
- They may prefer to spend their time with postdocs who, having already been trained in the fundamentals of the area, will (at least initially) be more mature and more productive.
- They are likely to spend more time being managers and grant-seekers and less time being scientists than used to be the case.

These downsides are not necessarily reasons to avoid working with the most prominent scientists, but they suggest that caution and careful investigation are in order. You have presumably done some of this in choosing this graduate school. Now you are in a position to do more at close range.

CHOOSE AN ASSISTANT PROFESSOR?

Could a junior faculty member be one of the leading researchers in the field and thus an appropriate choice for an advisor? The answer could well be yes. Hiring of faculty at the top research universities is so competitive that only the best graduates or postdocs from the best labs have a chance of being hired as assistant professors. They will have trained in labs with the highest standards, working on cutting-edge problems (which they generally bring with them to their new positions). They will have met most of the leaders in their field (a big reason to pick a program that brings in lots of visiting speakers), and these senior figures will know them from their publications and presentations at national meetings. They will be fully aware of the most important problems in the field and may well have already made a pioneering breakthrough or applied a powerful new technology. If you choose carefully, you will not be violating the rule of choosing the leading faculty by selecting an assistant professor.

Choosing an assistant professor as your research advisor can have many benefits. He or she will be full of energy, enthusiasm, and fresh ideas. Faculty members at this junior rank will most likely still be working at the bench with small numbers of students, rather than spending most of their time administering large research groups, so you'll get the benefit of direct contact and training from a top-flight scientist, rather than finding yourself under the supervision of a postdoc or senior graduate student. There will likely be greater opportunity for informal conversations on research and other topics than with a more senior, busier, and more removed faculty mentor.

Of course, there are also risks. You will not have as many coworkers to turn to if a problem arises and your advisor is not available. The lab may not be as well equipped as that of a more established scientist. (On the other hand, the most modern equipment may have been part of the assistant professor's start-up package, while an older professor's lab may be getting a bit musty.) If the assistant professor doesn't yet have substantial external grant funding, you may have to spend more time as a teaching assistant, rather than being fully supported as a research assistant. Assistant professors, although closer in age to their graduate students, may not be as experienced as mentors and may be too busy developing their own careers to spend as much time as desirable helping students. And of course there's the biggest risk of all, that an assistant professor may not get tenure, perhaps for failing to establish a suitably prominent and well-funded independent research program. In that case, the faculty member will have to leave the university, and you may have to find a new advisor if you can't graduate before he or she leaves.

All in all, the benefits of choosing an assistant professor as your research advisor are generally worth the risks, so long as you investigate carefully before making your choice. You will want to be particularly attentive to signs that your potential advisor may not get tenure. Establish a clear line of communication with your advisor that enables you to inquire periodically about progress toward tenure. Monitor overall productivity in the lab, including the rate of publication and status of grant applications, and compare these with the records of others who have—and haven't—gotten tenure in the department.

To help buffer against a young advisor's inexperience, however, it will be especially important for you to have other mentors. We'll talk about how to find additional mentors later in this chapter.

Initiating contact with potential advisors

Each graduate program will have its own way of putting new students in contact with prospective research advisors. There may be a series of faculty talks, a social hour, or a set of small group meetings. In any case, you'll eventually have one-on-one meetings with the few faculty members in whom you're most interested. These are obviously very important meetings, for reciprocal sizing-up as well as for garnering information. You may find they provoke anxiety, and we can testify from being on the other side that it's awkward for a faculty member to try to engage in conversation with a student who is nervous and ill-prepared, and who may be there only because the department demands that a student visit with some minimum number of faculty before indicating a preference for research advisor. At their best, however, such meetings can be lively and mutually instructive, so it's in your interest to prepare thoughtfully. Even if you don't end up working with a given faculty member, a good impression can be helpful as your graduate career progresses.

Your initial conversation will be purely provisional. You're not implying any commitment to work with this faculty member, nor is the faculty member committing to being your advisor. You're both there to learn about each other. In the next section we'll list a lot of questions you might ask a potential advisor. Here, let us list some of the things he or she might want to know about you.

What are your scientific interests? How do you think they connect with the work of this research group? Where did you do your undergraduate work, and with whom did you study? Did you do undergraduate research? If so, what was it about? Did you publish the work? Have you worked in the private sector, for example as a technician, in a job relevant to your intended future career? What special experience or skills do you have? Computing expertise? Familiarity with some unusual instrumentation or system? Field-work? Have you done anything unusual or interesting? Traveled to strange places? Lived overseas? Held an interesting job? Pursued an unusual hobby? What are your present career goals?

If there's something you'd particularly like the professor to know about you, be proactive in raising it.

Questions for potential advisors

Here is a long list of questions you might ask a potential advisor. Obviously, you won't cover all of these issues at your first meeting, and you may have other questions specific to your circumstances, but the list should give you some ideas to get started. You might also want to ask the students in the professor's research group about some of these matters, to get a perspective that may more closely reflect your own concerns.

Research projects. What is the general theme of research in the group? Are there any new directions that you might not be aware of? What specific research projects might you get involved in? Does the professor have a research grant that will support you if you work on these projects? Does the professor expect you to work on one of the ongoing projects in their laboratory, or to devise your own? Is there a gradual movement into full-fledged research, perhaps starting with a trial project, or are you expected immediately to begin the project that will be your thesis? Is the work experimental, theoretical, or both? Will you have adequate access to specialized equipment or other critical resources? Will that access be during regular working hours, or only on weekends or at night? (It's OK to work odd hours—it's part of the expectations and boot camp aspects of life as a graduate student—but you should be aware of the situation, particularly if you have family responsibilities that may interfere.)

Availability of a position in the research group. How many other students, postdocs, and technicians are already in the group? Does this seem like a good size? Is the professor interested in adding a new student to the group? (You should consider whether you want or need the individual attention that you'll get in a small group, or if you will be comfortable in a large group.)

Availability of funding. Does the professor have a research grant that will support you as a research assistant on your thesis project? How long will this grant last before its next competitive renewal? Does the professor expect you to apply for your own external funding? If so, what kind of help will you be given in preparing the application? Is financial support available for a non–thesis-related research assistantship, or for a teaching assistantship? How much teaching does the department and the professor expect you to do?

Availability of the advisor. Is the professor planning to spend a significant amount of time away from the university in the next couple of years (for example, on a sabbatical or extended research project)? If so, how will the advisor stay in touch with the group? Will some other faculty member, or a senior grad student or postdoc, be available for day-to-day supervision and advice? Does the professor have tenure? If the advisor is an assistant professor, what are the odds that she or he will be around until the completion of your degree? Is the professor close to retirement?

Course work versus research. Does the professor expect you to take courses beyond those required by your graduate program? Will the advisor be supportive if you want to take extra courses, or are you expected to take the absolute minimum so that you can spend as much time as possible on your research?

Publishing. Does the professor have specific expectations about how many research articles a grad student should publish while completing their thesis work? Does the professor mainly edit drafts prepared by the student, or do most of the writing based on the raw data? Does the professor expect students to write up their own work more or less independently, or is there extensive coauthorship with other lab members? If the latter, how is the order of authorship decided?

Questions for Current Students

Here are some questions, in addition to those in the preceding list, that you'll want to explore with students and postdocs already in the research group, to get a sense of how things really work.

Working conditions and morale. How often does the advisor meet with individual members of the group? Is there a regular group research meeting and journal club? Does the professor supervise graduate students directly, or are these duties delegated to senior graduate students or postdoctorals? How many hours a week are you expected to work? Is the professor comfortable with your coming in late and staying late, or are you expected to follow an eight to five schedule? Are vacations encouraged? Tolerated? Discouraged or forbidden? Is the professor always in the lab, checking on how things are going and offering help? Or is the style more hands-off, allowing independence and expecting only periodic reports? Does the group have good morale and a cooperative spirit?

Meetings and networking. How frequently do the group members go to professional meetings? Once a year? Once in a graduate career? Whenever there is something to present? Are travel and meeting expenses fully paid, or is the student expected to pay part out of personal funds? Does the professor help students network at meetings? Are visiting scientists invited to meet with the research group, or are they closeted with the professor?

Working with students and career development. Is the professor friendly and collegial in dealing with students? Businesslike? Hard-driving, demanding, or even abusive? Neglectful? How long do students in the lab typically take to get their degrees? Is there an unusual number of students who haven't completed their degrees? If so, why? Have students left the advisor and sought another? What kinds of jobs does research in this group prepare you for? What kinds of positions have recent graduates obtained? Is the professor supportive of students who are interested in a nonacademic career after receiving their degrees?

Questions for Yourself

Do you feel that you would enjoy working with this professor? Does he or she have a style that you would find congenial, or at least acceptable? Do you anticipate any problems due to incompatibilities of professional or personal styles? If you have, or are planning to have, a family, is the professor supportive of this? Does the lab offer maternity/paternity leave?

Be open to surprises

Choosing a graduate school and a research advisor are two of the most important things you will do in your professional life, but they are not the final decisions you will make about the direction of your intellectual interests. You may initially have several interests or labs between which you can't decide, and your first year or two in the graduate program will give you more information (both objective and relative to your own psychological bent) on which to base a choice. A few graduate-level courses or a couple of lab rotations—trying out different projects for a few weeks or months—may help you decide, for example, whether you prefer a more quantitative/ theoretical or a more descriptive/experimental approach. Or you may find something that had not even been on your initial list of choices unexpectedly fascinating. It's sensible to plan and choose as systematically as possible, but you should leave yourself open to the unexpected. After all, one thing that's predictable is that you'll be doing something quite different five years from now, and something else five years later. Make a rational choice, then go with the flow.

What to expect from your advisor

Now that you've chosen a research advisor, what should you expect from him or her?

BEING THERE WHEN NEEDED

First, your advisor should be there when needed. This does not mean always being on call, but he or she should be reasonably prompt and responsive when an issue arises. Some students have regularly scheduled meetings with their mentors, and if these take place every week or two, a student can generally wait for the next regular meeting. If, however, meetings are scheduled only once a month, or every six months, or if urgent advice is needed, then the mentor should be accessible on an ad hoc basis. On the rare occasions when an urgent issue arises, a student should be able to schedule an appointment, or at the least a phone or e-mail exchange, within a couple of days. We find that availability and rapid responsiveness are much appreciated by our students.

Not all of "being there" will be related to research issues. Sometimes you may want your advisor to listen and give advice regarding personal problems, doubts about your career path, or other matters. You have the right to expect a reasonable amount of patience and sensitivity at such times, along with full attention and no interruptions from phone calls or other visitors. It may be, though, that another mentor will be a more suitable or comfortable resource in such situations, or that you will need professional help or advice. Bear in mind that universities have counseling offices that can be helpful when you need confidential and neutral advice about academic stress, personal relationships, financial pressures, and similar matters.

KNOWING THE FIELD

Second, your research advisor should have a command of the major issues and problems in the field, including a knowledge of which approaches have failed and which remain promising, who the influential researchers are and who is doing the best work, which journals are publishing the most significant articles, and what meetings are most important to attend. It has been suggested that in the humanities, where the student chooses and works on a dissertation topic more independently, the advisor need not be a subject matter expert, but that's certainly not true in the sciences. If for no other reason, your advisor must be an expert to get the grants that will support the research and pay your salary as a research assistant. Of course, no advisor can be an expert on everything, so an important aspect of his or her expertise is knowing when someone else should be consulted for help with a particular technique, theoretical interpretation, or statistical analysis, and knowing who are the best people to contact.

PROVIDING CONSTRUCTIVE CRITICISM

Third, you should expect your advisor to be critical in a constructive manner. The advisor will suggest a research project and perhaps some relevant references and an angle of approach; the student then begins carrying out experiments, reports on his or her progress at research meetings, and drafts a preliminary account of the findings. At this point, the advisor should behave somewhat like an editor, critiquing the student's analysis and raising pertinent questions: Did you even consider this? What's your evidence for that? Do you think you should explore this other thing? As the project progresses and the student gains expertise and confidence, this feedback may become less detailed, but throughout the project, the advisor should remain available to troubleshoot problems and to offer ideas when research has stalled. At the end, there will be a serious and thorough critique in which both the evidence and its presentation (writing, illustrations, references, etc.) are thoroughly evaluated, by the advisor and by a dissertation committee.

HELPING TO PLAN FOR THE NEXT PHASE

Fourth, the advisor, along with other mentors, should help you plan for the next stage of your career. They should help you to clarify both short- and long-term goals. They should help to line up a postdoctoral position in a good lab, keep an eye out for suitable jobs, introduce you to influential and interesting people at professional meetings, let you meet with visiting seminar speakers, and help you to develop a professional network. They should provide advice on the timing of looking for the next position: not too soon, before enough is accomplished; but also not too late, which might result in wasted time. The advisor should allow you time, when you are ready to look for the next position, both to begin reading outside of your narrow research area to identify interesting projects and the people doing the best work, and to travel for interviews or to professional meetings.

JUDGING WHEN IT'S TIME TO STOP

It's always difficult to decide when a student has done enough work and should be encouraged to stop the research and write it up. This is a fifth responsibility of the advisor: to judge when enough is enough. There should be enough to produce several good journal articles, so that the capabilities of the student are clearly on display and the project is brought to some reasonable stopping point.

But the advisor should avoid keeping students on too long, taking advantage of them to obtain additional results that confer no additional benefit to the student. This is not an easy line to draw, since students may well benefit from researching and publishing a couple of extra papers after they have their techniques and methods and experimental system or suite of computer programs all polished up and operating efficiently.

It's for this reason that there should be a dissertation committee, to oversee both the student and the student-advisor relationship. The members of this committee should be viewed, by themselves and by the student, as additional mentors.

WRITING LETTERS OF RECOMMENDATION

One final responsibility of mentorship is to write letters of recommendation for the next job. Such letters have come to be viewed with considerable suspicion, as not being sufficiently frank about the deficiencies of the candidate. In our litigious environment, committing negative words to paper, especially words that may cost someone a desired position, can lead to a lawsuit. Writers of recommendations have become very crafty about avoiding the negative areas, or alluding to them only vaguely, and recipients of such letters have become equally savvy about reading between the lines. Our advice to mentors is generally to be honest about their degree of enthusiasm when asked to write a recommendation. What would be more constructive and helpful, however, is to identify areas of weakness while there's still time to do something about them, and to be straightforward in telling the student and suggesting ways to address them.

Overall, your advisor's job is to help you prepare yourself for an independent, rewarding career and professional life. An advisor should help you become confident and self-sufficient, not try to make you into a duplicate of herself through dictatorial or overly controlling behavior.

If problems arise

Sometimes problems may arise outside the lab that seriously impede your progress. A new situation in your own life may disrupt your planned graduate schedule: for example, the birth of a child, illness, or a family problem. If this happens, contact your advisor promptly and work out a new schedule. Be as realistic as you can, so that you don't miss a second set of deadlines.

At the same time, your advisor may get sick, have a busier than anticipated teaching or travel schedule, take on new responsibilities, or just become inexplicably neglectful and unresponsive. If this occurs, the first step is to remind the faculty member politely about your need for steady progress and feedback. Try to arrange a meeting in person, rather than by e-mail or phone.

If that doesn't suffice, you can discreetly ask other students whether this is normal behavior for the professor, what the program norms are, and whether the delay you're experiencing is something that you should be concerned about. They may have experienced similar situations and have suggestions about how to move things along. Other faculty can also offer useful advice. If you have another faculty mentor, he or she can offer suggestions and perhaps (if relatively senior in rank) intercede on your behalf.

If you need further assistance, it is advisable to follow the formal chain of command. You should next talk with the director of graduate studies and then perhaps with the department chair. If none of these have the desired influence, you should probably talk with someone in the graduate school. Staff members such as graduate program secretaries may be able to direct you to people and offices in the department, college, or university that can be helpful.

Changing advisors

If you're already in graduate school, working in someone's lab, and something goes sour, is it too late to change your decision? Well, maybe, or maybe not. If within the first year or two you find that things are not going well—that you don't like your project or your labmates or your advisor—you should think seriously about making a change. You've probably spent most of your time so far taking general courses that will be of use wherever you end up, and those efforts will not have been wasted. Beyond your second year, however, switching projects will be costly; you will have started to accumulate results, which most likely you will not be able to use in a new thesis project. Most graduate programs are sympathetic to a student who seems bright, competent, and motivated but is looking for a better fit. Most professors will understand too, since they are probably also aware that things aren't going well. After all, it's your career and your life. Give it serious thought, and don't make a spur-of-the-moment decision, but make the decision that seems right.

If you are considering making a change, then you might talk (discreetly) to other students, to another faculty mentor if you have one, and definitely to the director of graduate studies and to your potential new research advisor. Be sure to follow any applicable rules and protocols set out in your school's policy handbooks. At some point, of course, you'll want to talk with your current advisor, and it's important that you do so in a way that does not cause bad feelings between old and new advisors. If it's a matter of not feeling the research project is right for you, despite good relations with the advisor, then you can have an up-front discussion. If, on the other hand, you want to leave the group because you and the advisor don't get along, then you'd be wise to have a secure landing place before you announce your departure.

Of course, if you are experiencing sexual harassment or other abusive treatment from your advisor, you should let someone in authority know as soon as possible. Most universities have a conflict resolution office or an ombudsman that can be useful in such situations.

Changing department or university

You may also find that the department you're in isn't quite right for you. Say, for example, you enter graduate school in physics, develop an interest in the applications of physics to biology, and find that your department has no biophysics researchers and no interest in working out a joint thesis project with a professor in biophysics or biochemistry. In that case, you might consider transferring to another department that shares your research interests. Talk to the relevant staff person in the graduate school about how to accomplish this. Such a transfer will probably require application and admission to the new program and will most likely mean losing any financial aid provided by the program you're leaving, but it should be feasible early in your graduate career.

Sometimes you'll be tempted to change not just your advisor or department but your university. This is usually not advisable, since the new school might wonder what's wrong with you, not what's wrong with the department you want to leave. Taking a definite step up, from a department generally viewed as mediocre to one with a top-notch reputation, is a different story. If you have very good recommendations both from your undergraduate school and from the department you're leaving you may be able to win transfer and admission, but don't count on it.

One situation in which you might consider changing universities is if your advisor accepts a position at another university. Whether it is feasible to follow your advisor to a new school depends largely on what stage you're at in your graduate career. Early on, before you've taken many courses or preliminary exams, the process should be relatively straightforward. The faculty and administration at the new university will be anxious to make your advisor happy, and happiness includes accommodating his or her existing students. Things get harder if you're further along; the courses and qualifying exams you've already taken may not conform to the new university's requirements. Special negotiations may be required, which—if you're lucky—your advisor will help you with. If you've reached the all-but-dissertation phase of your graduate career, just doing research and writing your thesis, you'll probably be better off staying at your original university and working with your advisor long-distance to finish your degree. Your director of graduate studies should be able to help with the necessary arrangements.

Other mentors

Your research advisor cannot and should not be your only source of advice, contacts, and support. He or she may not always be around, will often be tied up with the manifold responsibilities of a faculty member, and may not be familiar with all of the details of campus or departmental life about which you need advice. Sometimes, too, you may need counsel because you're having trouble with your advisor. It's important that other faculty members know who you are, so you can call on them for advice and for letters of reference. In other words, you need a network of mentors, not all of whom have to be faculty members.

MENTORING COMMITTEES

It's advantageous to have faculty members other than your research advisor whom you can call on for advice. You might start with the director of graduate studies, who may be the first faculty member you talk with at length when you join the program. But as you get deeper into your research, the most valuable will be your dissertation committee: a group of two or three faculty members, in addition to your advisor, who meet with you periodically to hear about your research progress and problems and to give advice about rate of progress, ways of avoiding pitfalls, and new approaches. The people on such a committee will usually be professors with whom you've taken courses or had some research interaction. Some may be more knowledgeable than your advisor regarding particular techniques or resources, which is important in these days of increasingly interdisciplinary research. The committee will have a chair, often not the research advisor, with whom you should periodically discuss your progress.

Committee members can also serve as a buffer between you and your advisor if there are areas of disagreement—such as whether you've done enough work to begin writing up your thesis or whether a particular line of research is likely to be productive. They can serve as references when you're looking for your next job or postdoctoral opportunity, informing you of opportunities and putting you in touch with contacts. See chapter 23 for more information about dissertation committees.

You may also meet a range of faculty members in orientation sessions held soon after your arrival at graduate school. And you may seek advice from the professors teaching your classes, if either the subject matter or their personalities seems appealing. These interactions will generally be based on professional suitability rather than personal friendship, but the two are not mutually exclusive. You might want to include both senior and junior faculty in your circle of mentors: the senior for their experience and contacts, the junior for their better appreciation of the trials and tribulations of young scientists.

SENIOR GRADUATE STUDENTS

Other students can be among your most valuable sources of advice. Senior graduate students can give you insights into the various faculty members in the department, how they work with their students, and how their research groups run. They can alert you to political issues in the department that you should not run afoul of. They can advise you about how to negotiate the bureaucracy. They can also help you get involved in social activities and departmental student organizations. Don't rely on just one senior student, no matter how charismatic, as he or she might have a jaundiced or biased view; get your advice and insights from a broad group of students.

DEPARTMENTAL STAFF

In many departments and graduate programs, it's a secretary or administrator who holds things together on a day-to-day basis. This may be the person with whom you correspond when you apply for admission and arrange your recruiting visit, or it may be a more senior person who works behind the scenes. In any case, these are often the people who know how the place really works, whom to talk to, and how to get things done. They're usually very interested in helping students. You should definitely be on good terms with these professionals.

YOUR MENTORING NETWORK

In addition to faculty members, other students, and staff in your own program, you may want to make contact and seek advice from faculty and students in other departments, alumni, professionals in the community, and scientists at other institutions. And don't forget faculty from your undergraduate institution. All of these people can give you useful advice, support, and career counsel. You will need to be proactive in assembling and maintaining your network of mentors—it won't assemble spontaneously—but it can be of enormous help to you in graduate school and throughout your career.

Personal relations with your advisor and faculty mentors

Handling personal relations with your advisor and faculty mentors is an important and potentially tricky aspect of graduate life. The basic rule is to be professional. Friendship may come and is potentially a nice addition, not to be refused, but there are boundaries that should be crossed only with care. Even in a friendly relationship between a faculty member and a student, there's a hierarchy involved: the professor has the unquestioned right, indeed the duty, to give constructive criticism to the student.

You should certainly not be expected to do personal chores for your advisor: babysitting, lawn mowing, helping to paint the house. Sometimes the advisor will enlist research group members to help with, for example, a move. This can produce camaraderie within the group, and it is not necessarily improper; but you should always feel free to choose not to participate, and should never feel coerced.

In academia it has not been uncommon for faculty members and students to become intimately involved. It still happens, but it's frowned upon these days. At many schools, institutional regulations prohibit or at least strongly discourage such relations because of the inherent power differential. As a result, other faculty members tend to disapprove. Other students may also be resentful of the perceived or actual favoritism that results from such relations. The principals themselves may, if the relationship sours, find continued professional interactions to be awkward, if not impossible. A student's graduate career can be derailed by a broken relationship with a professorial advisor, and a professor's career can be damaged by a grievance or lawsuit. The bottom line is that you should enter into such relationships with extreme caution, and preferably not at all, at least until the teacher-student relationship is finished: the course is completed, the thesis is written and approved, the articles are published. When such restraint is not possible, changes in supervision and evaluation should be arranged. Some faculty-student relationships have happy outcomes, but they may be in the minority. Bear in mind that you have significant control in fending off unwelcome romantic or sexual advances, particularly in their early stages, before the situation gets more complicated. Use available institutional legal and other counsel if needed.

At a less intense level, what about friendship? This is most likely to come about between a younger faculty member and the students in the department. They're of similar age; the faculty member remembers what it was like to be a grad student and may not yet have family obligations. Having parties or picnics or going bowling can be a welcome break from the grind, a reminder that there are other things in life besides work and studying, and a way of deepening and diversifying friendships with compatriots in a professional discipline. They can contribute substantially to esprit de corps and morale within a program: students get a chance to interact on a roughly equal basis with faculty members, while faculty members can feel that they haven't yet lost their youth.

Such interactions can be a very good thing, so long as they don't get too personal. But they are not an essential part of the mentoring relationship, which should be based primarily on professional advice and counsel, offered from a position of greater experience and knowledge (which one hopes doesn't translate into pomposity). There are some faculty members who limit their social interactions with students to a departmental picnic or Thanksgiving dinner for students far from home, and who serve very well as effective mentors. There are some pressures to be chummy, regardless of age or status difference; but faculty mentors do well to resist pressures to be either your parent or your best friend.

Take-home messages
- Consult with more advanced graduate students in the program to find out how they enjoy working with their advisors.
- Take advantage of first-year lab rotations, if they are available in your program, to obtain firsthand knowledge of what various potential advisors and their research groups have to offer.
- Choose an advisor who is not only a good scientist but also a good person. This combination will provide you with both an advisor and a mentor.
- Select an advisor who will help you attain your future career goals.
- Consider the makeup of your advisor's research group to ensure personal and professional support.

References and resources

Adviser, Teacher, Role Model, Friend: On Being a Mentor to Students in Science and Engineering. 1997. Washington, DC: National Academy Press. http://www.nap.edu/readingroom/books/mentor/.

Green, Anna L., and V. LeKita, eds. 2003. *Journey to the Ph.D.: How to Navigate the Process as African Americans.* Sterling, VA: Stylus Publishing.

Johnson, W. Brad, and Jennifer M. Huwe. 2003. *Getting Mentored in Graduate School.* Washington, DC: American Psychological Association.

Oliver, J. E. 1991. *The Incomplete Guide to the Art of Discovery.* New York: Columbia University Press.

Rackham Graduate School, University of Michigan. 2006. "How to Mentor Graduate Students: A Guide for Faculty at a Diverse University." http://www.rackham.umich.edu/StudentInfo/Publications/FacultyMentoring/Fmentor.pdf.

Zelditch, M. 1990. "Mentor Roles." Proceedings of the 32nd Annual Meeting of the Western Association of Graduate Schools, 11. Tempe, AZ, March 16–18.

4

CHOOSING AND CONDUCTING A DISSERTATION PROJECT

Your goal in graduate school is to do research that will lead to publications—including a dissertation—that will make a significant contribution to scientific knowledge and establish you as an expert in your field. Note the difference from undergraduate school, where the goal was to take varied classes, study hard, get good grades, and learn about a wide range of things. Of course, you will still take courses and learn many new things, but all will be directed toward your primary goal: to make your mark in your discipline. This chapter discusses how to choose a research project that will advance this goal.

A good research project is one that addresses an important issue in the field, is feasible but not trivial, and is engaging enough to sustain and excite you through several years of hard work. Identifying such a project is difficult—indeed, it is often said that the mark of an exceptional scientist is the ability to choose the right problems. As a beginner in your field, you may well ask how you can find a good project. We can't guarantee success, but the ideas in this chapter show how you can undertake the search systematically and maximize your chances.

It's important to recognize, however, that finding a good research project cannot be totally systematized. The field, the feasibility of certain approaches, and your own interests can change unexpectedly, so the project you end up with may not be one you arrived at by way of an initial rational search. Nonetheless, as Louis Pasteur said, "In the field of observation chance favors only those minds which are prepared." So start out on a rational path, and see where it leads.

We have written this chapter with the assumption that you will have a major role in defining your research project. In reality, your advisor may assign a project, particularly if you are being supported by one of his or her research grants. Even then, however, the advice we give—about how you can actively participate in the development of the project ideas and implementation—should remain pertinent.

Get familiar with the literature

A key part of choosing a project, once you've decided to work with a particular advisor, is to become familiar with the relevant research literature. Reading to choose or prepare for a project is an important activity, one that is often not given as much attention as it deserves. But as one wit put it, "Two months in the laboratory can save you a week in the library." You may start out with textbook material, then review articles, then primary journal articles from your advisor's laboratory and the labs of the main competition. You should learn what the major journals in the field are and read (or at least scan) them regularly. As soon as possible, you should begin to use current literature, abstracts, and if possible meetings and personal contacts to see what's going on now in the field.

As you narrow down the research area within which you will develop a specific project, you should try to get a clear sense of what is reliably known and what is uncertain, shaky, or simply unexplored. As you read you should familiarize yourself with the techniques, materials, and methods typically used in your field; but you should also be thinking about whether more creative approaches might be productive and might give you insights that others have not had. If there is a strong theoretical basis to the field, you'll want to spend a fair bit of time reading, working problems, and perhaps becoming familiar with or writing computer programs. If statistical methodology is important, that may also require special study.

It's important to emphasize, however, that the goal of your research is original results, not total familiarity with the literature. Getting to know the literature is essential, but don't overdo it and try to read everything. Striking a balance is the key. It's also worth noting that you should not decide to pursue a question simply because it has not been addressed before—there might be a good reason for that. It's possible that the question cannot be adequately answered with available techniques, instruments, and reagents. More importantly, the question might be too trivial to have attracted the attention of others.

Attack an important problem, and be able to articulate its significance

Perhaps the most crucial thing you can do when choosing a research project is to pick a problem whose solution will make a difference, a project that affords you a realistic chance of making a contribution during your three to five years of thesis research, but also one whose outcome will matter to others. Every field has more than one important problem, so you need never fear that you and all your contemporaries are racing to attain the same goal, all competing for a single prize.

You should begin with your own understanding of what's important. You presumably decided to concentrate on your chosen field of science, at least in part, because you were intrigued with the questions being posed and the potential impact of the answers. Now is the time to deepen your understanding of these issues. Since you are relatively inexperienced, your judgments may not be entirely realistic, but you have the advantages of youthful idealism and a fresh vision, which more established scientists may lack.

Start by asking what problem you will be trying to solve. Not in the narrow sense of solving for the unknown in an algebraic equation, but in the broader sense of identifying some substantial gap in our understanding of nature. Even if your potential thesis project will not fill the whole gap, you should feel that it could make a substantial contribution.

At the same time, try to put the work in a broad scientific context. Any particular scientific project should be part of a larger fabric, with broad and lasting consequences. It should have implications for both closely related and more distant fields. Ideally, your project should have the potential to become the basis for work by others. You may have only vague ideas at this point as to what its fruits may be, but you should aspire to this sort of fertility.

You should also be able to explain to nonspecialists what the project is about and why it's important. What will this work do for people, if you are successful with it? Will it improve their health or material well-being, advance their intellectual development, or deepen their understanding of themselves and others? Being able to articulate this general understanding is important for two reasons. First, we live in a world with other people, who to some extent support our research and need its findings. It will help us as well as them if we all understand what we are doing and why. Second, scientists spend a lot of time on their work but also interact socially. Researchers who can talk engagingly about what most engrosses them will be better able to maneuver in public settings, a source of satisfaction that will complement and build on those from the laboratory.

It is worth noting that the National Science Foundation has two merit review criteria for research proposals: (1) What is the intellectual merit of the proposed activity? (2) What are its broader impacts? Being able to persuasively address the broader significance of your project will thus increase the chance of having a proposal funded, while at the same time making your work more intelligible to nonspecialists. For examples of potential benefits that may satisfy the broader-impacts criterion, see http://www.nsf.gov/pubs/gpg/broaderimpacts.pdf.

Pick a problem that excites and fits you

You will be working on the problem you choose for several years, so it should be something that engages you emotionally and aesthetically as well as intellectually. The particular choice will depend on your field. If you're an ecologist, for example, you might choose between studying birds or sea turtles based on a longtime love of bird-watching or a desire to spend more time on a tropical beach. In a neuroscience program, it might be a matter of choosing between neural networks and nerve growth factors depending on whether you are hooked on computer programming or especially enjoy wet lab work.

Your project should also fit your working style and intellectual preferences: Would you prefer a project that is fully experimental and rather empirical, experimental but with good theoretical underpinning, or mainly theoretical or computational.

Pick a project that is feasible

It is important that your project be significant and exciting, but it should also be one that you can actually complete in two to three years of research, and one that is likely to generate a reasonable number of papers in high-quality journals.

Nobel Prize–winning economist Herbert Simon has written, "The quality of a research problem rests on the importance of the ideas it addresses and the availability of ideas and techniques that hold out a promise of progress" (1991, 366). There are many projects that seem very attractive, especially to newcomers to a field, but that are simply infeasible due to inadequate analytical methods, insufficient supplies of pure material, an excessively complex or murky theoretical background, or the like.

Eventually, some of these obstacles may be overcome. Indeed, it is common in science for a field to move rapidly for a decade or so, stall when available techniques are no longer adequate to solve the remaining problems, and then revive two or three decades later when a new, more powerful set of tools or ideas has been developed, often in a quite different field. Sometimes only modest improvements in technical capabilities are required (e.g., an order of magnitude or less in yield or sensitivity), and working on such improvements might reasonably be part of your project. In other cases, however, much larger steps are needed. You will then have to decide whether to devote your PhD career mainly to developing new techniques, rather than solving a physical or biological problem. Either choice is reasonable, and many of the most cited and influential papers address technique. But you should recognize the choice you would be making and be sure it's one you want to make.

Get your feet wet

Having chosen a research advisor and a research problem that seems important, exciting, and feasible, you need to jump in and get your feet wet. In the first year or two of graduate school you probably won't have time to do much more than that. Much of your time will be spent taking classes, being a teaching assistant, and preparing for preliminary exams. You should, however, spend as much time as possible learning about your research project: reading the literature, learning to work with the instruments particular to this area, and getting practice with simpler projects before embarking into the fully unknown. Often the first large block of time available to work on the project is the summer after your first year in graduate school.

The most practical way to proceed is usually to begin work on some corner of a bigger problem. This first step will likely be defined by your research advisor, modified only slightly in response to your own taste and knowledge. By working on the technical details until you can either reproduce existing results or show that something's wrong with them, you will gain expertise and insight and experience. Keep an eye out for new techniques that might produce the desired results more quickly or reliably or with a greater degree of sensitivity. Read the basic literature dealing with the problem, and with related problems, to get a sense of where your problem fits in the bigger scheme of things and how new and existing techniques and approaches might be applied to it. But don't just read: talk to people about your project and ideas—grad students in other labs, visitors, friends of friends with whom you're having an evening out—and go to departmental lectures where you might hear something pertinent.

Your purpose at this stage is not to embark full tilt on your dissertation project, but rather to get a sense of the realities of the area, to get some knowledge into your hands by doing experiments or typing computer code, and into your head by reading journal articles. Such "tacit knowledge" (to use Michael Polyani's phrase) will give you a much better sense of what's easy, what's essentially impossible, and what's both feasible and important.

Start with your advisor's idea, and make it better

Once you've gotten your feet wet, you're in a position to begin thinking critically about your project. One of the distinguishing features of basic research is that it cannot be fully planned. New leads and new opportunities arise unexpectedly. You should spend at least a small part of your time wondering how you could make your project better.

Your project should reflect a balance between your independent thinking and your advisor's knowledge and expertise. Your advisor has probably been working in the area for some time, undoubtedly has ideas about what the most promising new directions are, and may have suggested an approach to your project. If he or she is a good scientist, these ideas should not be cast aside lightly. But even a good researcher, having worked in a field for a long time, becomes a bit routinized. There may be a better approach, and your advisor should at least be willing to consider other possibilities, including the use of new techniques.

New techniques—experimental, theoretical, or computational—often come from other fields. You may be lucky enough to recognize an opportunity that your mentor (who has been productively plowing the ground with existing techniques) has overlooked. This is one of the best ways to improve an existing project. If the new technique is not overly elaborate or expensive to implement, you might suggest deploying it in your advisor's lab, perhaps after trying it out on a couple of typical cases in a lab (in your department or elsewhere) that already has expertise. If the technique is very expensive, requiring specialized instrumentation, then perhaps collaboration between your lab and that of another professor could be arranged. By acting as the go-between you may gain expertise in both labs' research areas.

Be aware that the most important thing to learn about a new research technique is its limitations: high background noise, for example, or susceptibility to interference from neighboring electronic equipment, the prevalence of false positive or false negative results, or sensitivity of reagents to light.

Working with your advisor

You need to work effectively with your advisor during your dissertation research, getting advice and feedback while gradually developing your independence. Here are some suggestions for getting maximum benefit from the relationship while using both your time and your advisor's time efficiently.

SET GOALS

Work with your advisor to establish short- and long-term goals and a timetable for reaching them. The long-term goals might be to get a PhD in five years and to line up a suitable job. The interim goals will include selecting and satisfactorily completing the classes needed for your degree program, preparing for and taking your written and oral preliminary exams, reaching specific milestones in your thesis research, and making contacts for your job search. Periodically (say, once a semester) meet with your advisor to review these goals, decide what must be done to reach them, and make changes if necessary.

Remember that, while your advisor has an interest in your being productive in research and writing, these goals are primarily yours. It's your education and your career; you must take responsibility for letting your advisor know what you're aiming for, what progress you think you're making, and what problems you're having. Your advisor—and other mentors—are there to help, not to be avoided.

HAVE REGULAR MEETINGS

Each advisor will have his or her own way of meeting with members of the research group. Some will meet with each student individually, some will have group meetings, some will have both. The advantage of one-on-one meetings is that the student gets individual attention, and topics can be addressed that might be awkward in a group setting. The advantage of group meetings is that everyone learns about all of the lab's research efforts and progress, all can contribute to solving a variety of problems, and students get practice presenting their research results to an audience. Having both individual and group meetings combines the advantages but at the cost of extra time.

Whichever system your advisor uses, be sure that you get regular advice and feedback. It is not wise to pursue a project for months without discussing progress with your advisor; you may find out later that you used incorrect experimental conditions or overlooked a crucial control necessary for data interpretation. The other extreme—asking your advisor daily what you should do next—is not healthy either. Graduate students often avoid frequent reports to their advisors because they want to work everything out on their own, or because they feel they should be able to operate independently, or because they're having trouble and are afraid to admit it. On the other hand, advisors are often busy and feel they don't have time for regular meetings, and some want to hear from students only when they're bearing good news, for example, that a manuscript is ready to be submitted. Whatever the reason, infrequent and irregular contact between student and advisor is not good for the student.

Regular meetings provide a useful rhythm, a predictable schedule for reporting progress and problems. The meetings should be spread out enough that there's time for things to happen, but not so infrequent that there's too much time to drift. Once a week to once a month is probably appropriate. Talk with your advisor about what seems sensible and feasible to both of you.

MAKE THE MEETINGS USEFUL

Come to each meeting with a list of things you'd like to discuss and accomplish: research progress since the last meeting, problems that have arisen, what you plan to accomplish before the next meeting. If you have decided on a certain approach, you should be prepared to justify the choice and to ask advice regarding specific experimental details. Better yet, propose multiple experimental approaches and discuss their pros and cons. This will enhance your training in asking proper scientific questions and designing the best experiments to answer them.

In an individual meeting, you can also take the opportunity to ask your advisor for a letter of recommendation, to suggest new supplies or equipment or software that you think would benefit the group's work, or to mention a professional meeting that you'd like to attend. Try to ensure that the most important items are covered first, to avoid running out of time. At the end of the meeting, try to summarize what has been accomplished and decided. Follow up with a written reminder, on paper or by e-mail, if your advisor has agreed to do something that might slip his or her mind.

BE FLEXIBLE BUT PERSISTENT

Sometimes you or your advisor will have to postpone or cut short a scheduled meeting because of unanticipated work or a personal emergency. Be sure that you let you advisor know if you can't make an appointment, and hope that he or she will do the same for you. Try to reschedule the meeting or arrange to communicate by other means. Conversely, when something comes up that needs immediate attention, don't waste time spinning your wheels if a quick response from your advisor would solve the problem. In either situation, it's useful to know what mode your advisor favors for impromptu communication: e-mail? phone call? a drop-in office visit? Is he or she willing to be called at home?

GET TIMELY FEEDBACK AND CRITICISM

When you've done something that calls for prompt feedback—such as completing a draft of an abstract for an upcoming meeting, a chapter of your thesis, or a fellowship proposal—make sure your advisor knows that the matter is time-sensitive, and ask whether he or she can get back to you in time. Allow as much lead time as possible, particularly if you know that your advisor's workload is especially heavy. Ask whether it will be OK to send a reminder a week or two ahead of time.

When the response comes back, accept any criticism in an objective manner. Be open to changes in wording, suggestions for clearer reasoning, and urgings that additional data and calculations or less speculation would improve the product. If after reflection you disagree with a criticism or suggestion, don't be shy about defending your point of view—but do it in a thoughtful, professional way.

INVOLVE MEMBERS OF YOUR DISSERTATION COMMITTEE

Work with your advisor to select the members of your dissertation committee as early as possible. You will want to discuss your progress with them periodically and to seek their advice regarding obstacles or competing possible directions. These discussions will familiarize them with your project and will give them the chance to give it their approval at successive stages of development.

ACT LIKE A PROFESSIONAL

Nothing is more rewarding to an advisor than to see a student progress from uncertain, naïve beginner to self-confident, knowledgeable researcher and junior colleague. You can demonstrate your growing professionalism by becoming increasingly conversant with the literature and major researchers in your field, by telling your advisor about relevant new work you've come across that he or she may not have seen, and by suggesting new techniques and approaches that you've encountered or thought of.

Read articles that your advisor suggests critically and comment on them thoughtfully. Show that you've embraced the idea that you are becoming the world's expert in your special area.

Restrictions on research

Not all research projects allow you to charge blithely ahead. If your research involves working with human or animal subjects, or with hazardous substances, there are many regulations you will have to be aware of and permissions you will have to gather.

Restrictions regarding human or animal experimental subjects are based on the fundamental principle of ensuring proper behavior toward living subjects. With human subjects, the key concept is informed consent. People must not be coerced into participating in experiments against their will, and they must be informed of the experiments' likely risks and consequences. Violations of these rules range from the outrageous and universally condemned (e.g., Nazi experimentation on Jewish and other inmates of concentration camps) to the more subtle and debatable (e.g., deceptive psychological experiments). All universities and research institutes have human subjects committees (institutional review boards, or IRBs) that are required to examine proposed experiments and decide whether they meet ethical guidelines.

There is considerable debate about experimentation with animals. While most scientists believe that experiments on animals are necessary to develop knowledge that may be applied to human disease, opposition from animal welfare and animal rights activists is increasingly strong. Even if the necessity for animal experimentation is upheld, it is important (both for fundamental ethical reasons and to deprive animal rights protestors of additional arguments) to treat animals properly. This includes not subjecting them to undue pain and anxiety; being sure they are properly fed, housed, and cleaned; and giving them proper exercise and socialization. Federal regulations in this area are becoming increasingly rigorous. Research institutions have committees that oversee animal research and can provide guidance.

If you are working with radioisotopes, hazardous chemicals, or biohazards, your responsibility is to use, store, and dispose of them so that you don't endanger yourself or others. Your institutional health and safety office will have instructions and training materials and will provide periodic training sessions, to which you should pay careful attention.

Take-home messages

- When selecting a research project, carefully review the relevant literature to find out what has already been done.
- Work closely with your advisor in deciding on a project to benefit from their experience and knowledge of the field.
- Aim for a research project that is both significant—addressing important gaps of knowledge in the field—and feasible.
- Choose a project that is fulfilling to you, one that will motivate you to put in the extra effort required for success.
- Discuss your progress regularly with your advisor and members of your dissertation committee.

References and resources

Parent, Elaine R., and Leslie R. Lewis. 2005. *The Academic Game: Psychological Strategies for Successfully Completing the Doctorate.* West Conshohocken, PA: Infinity Publishing.

Peters, Robert L. 1997. *Getting What You Came For: The Smart Student's Guide to Earning a Master's or Ph.D.* New York: Farrar, Straus and Giroux.

Phillips, Estelle M., and Derek S. Pugh. 2000. *How to Get a PhD: A Handbook for Students and Their Supervisors.* 3rd ed. Philadelphia: Open University Press.

Polanyi, Michael. 1974. *Personal Knowledge: Towards a Post-Critical Philosophy.* Chicago: University of Chicago Press.

Simon, Herbert A. 1991. *Models of My Life.* New York: Basic Books.

5 EFFECTIVE TEACHING

If you decide to seek employment at an academic institution, you will undoubtedly be asked to teach. Depending on the nature of the institution, you might teach undergraduate, professional, or graduate students. While it is customary for graduate students and postdoctoral trainees to learn from their advisors how to do research, their advisors rarely teach them how to teach. Some are innately excellent teachers who develop their own teaching strategies. Most, however, teach as they were taught. Of course, most students have encountered both good teachers and poor teachers. Thus, our most important recommendation for your training as a teacher is to reflect carefully on the teaching style of every instructor you have had. With your current teachers, take note of what works for you as a student and what doesn't, keeping in mind that the goal of teaching is learning. Pay attention to class content, organization, and dynamics. Observe the instructor's style of presentation, demeanor, and interaction with students. Reflect on what you would like to apply and what you would rather avoid as a teacher. Think of what you can improve and how you might capitalize on your personal strengths. You should also invest time in formulating your own teaching philosophy. This will serve as an invaluable guide for all details to come.

A teaching philosophy usually starts with your own concept of what constitutes effective teaching and learning. State the goals you have for your students and the skills you expect them to develop, and describe your style of interaction with students. Looking into the future, state your plans to continue growing as a more effective teacher. See chapter 9 for more details.

We do not know many people who have started a teaching career without some initial anxiety. It is natural to worry about what to include in a lecture, how to present material in a manner that encourages students to learn, and the best ways to add to and clarify the content of a textbook. You will have many questions on your mind: Will I remember what I want to say? Can I cover the material on my syllabus in the allotted time? Will I be able to answer the students' questions? How will I deal with problem students? How do I determine whether students are learning? Will students like and appreciate me as a teacher?

Your major allies in dealing with such anxiety issues are your fellow teachers. Talk to them about your worries. Ask what works for them and if they would share, for instance, examples of their syllabi, handouts, and tests. Many institutions also have teaching and learning centers where you can find experts formally trained in various aspects of pedagogy. Utilize these valuable resources to obtain advice about syllabus structure, course evaluation, and other matters. You might even ask for a teaching consultant to attend and observe some of your classes. This is a good way to get feedback to polish your teaching skills. Your students can also be helpful. When dealing with them, maintain open communication to find out what needs improvement. Explain your teaching philosophy and set out the decisions you have made about class format, grading, and so on.

Learning how to teach well is worth the effort. It will make teaching fun, not a dreaded chore, and will provide you with great satisfaction. Gaining teaching experience and learning about what makes an effective teacher while in graduate school will provide you with a head start during your first couple of years as a faculty member. More importantly, it may help you land that faculty position. Many institutions offer courses on teaching skills to graduate students. The best approach to improving your teaching skills, however, is to actually teach. For a start, ask your advisor to allow you to teach a part of a lab or small group session. Then ask your advisor for feedback. If you are both satisfied with your competency, seek other teaching opportunities. Remember, however, to consult with your advisor to ensure that these added responsibilities do not interfere with progress in your research.

You will soon find that there is no end to improving your teaching skills. By trial and error and sharing ideas with others, you will continually find new ways to achieve better student learning. (An invaluable key here is self-reflection following each class session. Write your thoughts down on paper; otherwise, specific memories of what worked and what did not will soon fade.) Another impetus should be the fact that knowledge in the sciences and engineering advances very rapidly and is becoming more interdisciplinary in nature. You will discover at some point that an approach to teaching that has worked well for you in the past does not work with a given group of students or a new subject matter.

Good teachers realize that not all students learn in a similar way. Some are active learners who absorb information through discussion, application, or explaining it to others, and some are reflective learners who prefer thinking about the presented material. Students can also be classified as sensing or intuitive learners. Sensing students have an easier time memorizing facts and solving problems using traditional approaches. Intuitive learners, by contrast, learn better by contemplating possibilities and connections; they prefer assimilating pieces of knowledge and are therefore less likely to be intimidated when answering questions based on material not explicitly presented in class. Students also vary in their background knowledge. This diversity of learning styles means that you will have to present material in a variety of ways, especially in the first few classes of a course.

Types of learning

Not all teaching results in learning, and not all learning is the same. Benjamin Bloom has classified learning into the following categories. The verbs in parentheses serve as a guide to the learning skills encompassed in each.

- Knowledge (define, repeat, list, recall, name)
- Comprehension (discuss, describe, explain)
- Application (use, demonstrate, practice)
- Analysis (compare, diagram, question, categorize)
- Synthesis (compose, design, arrange, assemble, construct)
- Evaluation (revise, select, assess, estimate)

Obviously, learning becomes deeper and more valuable as one progresses from simple gaining knowledge to becoming able to analyze and evaluate it. You should aim at teaching that results in these latter types of learning.

Establishing course goals and objectives

Well before you meet your students, you must think carefully about the goals and objectives of your course—what it is that you expect students to learn. Do you want them simply to acquire a set of facts (a bad idea!) or to step it up a notch, developing the ability to analyze and synthesize learned material and to move toward solving problems. Course objectives should fit the goals of your audience, so think about the educational background, academic stage, and career objectives of your students. Ideally, you would like to promote enough interest in the subject matter that students will be motivated to learn more on their own. Setting up course goals will also help you formulate a syllabus, choose a textbook, and decide on the most appropriate methods of teaching and assessing student learning. All of these choices influence the degree of success in achieving the course goals.

Selecting a textbook

Which textbook you choose for a given course will depend on many factors. You might be lucky and find a text whose contents perfectly fit the course goals. However, this is often not the case for higher-level undergraduate courses and even less so for graduate courses. Each textbook has unique strengths and weaknesses in terms of content and presentation, and, given the rapid pace of advancement in science, many are outdated soon after publication. Thus, whatever book you choose, you may need to supplement it with additional reading material.

Course syllabus

A syllabus is essentially the road map that will guide you and the students through a course. Unfortunately, some teachers do not spend sufficient time and energy developing an informative course syllabus. Their syllabi may include only the course number and title, the instructor's name, and the dates of lectures and their topics. A good syllabus, by contrast, should also contain information on how and when to contact the instructor; a brief description of the course, including its goals and objectives; a reminder of any prerequisites; lists of required books and other reading material; specific assignments; and statements on late assignments and incomplete course work, attendance and grading policies, and the expected class environment. It should also contain summaries of or references to policies on cheating, plagiarism, sexual harassment, and other aspects of student conduct. Some teachers round out their syllabi by including statements on diversity, accommodations for students with disabilities, and students' expectations of the teacher; advice on studying for quizzes and exams; or articulations of their teaching and grading philosophies.

The first class session

The first class session can set the tone for the entire course. Up to this point, neither you nor the students have known exactly what to expect. You might break the ice by telling students about yourself, both professionally and personally. Share with them your research interests and how they relate to the course topic. In small classes you might ask students to introduce themselves briefly, describing their backgrounds and telling why they are taking the course. You should always tell them what you expect them to learn and what role you will play in facilitating that learning. This is also an opportune time to ask them about their expectations of you. Discuss with them the course objectives, the overall structure of the course, and the highlights of the syllabus. (Do not read the entire syllabus to the students!) More importantly, you should share with students your teaching philosophy and method and explain why you have structured the course as you have and why you chose the particular textbook. Invite questions. One way of starting the conversation is to have students work in pairs to identify and paraphrase these elements in your syllabus. None of this is a waste of time. On the contrary, it will make your task easier and will make the entire course a more pleasurable experience, for you and for the students. This discussion will make the students feel you are there to help them learn rather than to intimidate them with your knowledge. The first class meeting is important, but do not be discouraged if it does not go as well as planned. There is ample time for recovery.

Giving a good lecture

A central attribute of good lectures is their clear connection to the course's goals and to past classes in the course. Thus, it is helpful to start each lecture with a summary of the previous one, and then to list the learning objectives of the new lecture. Learning objectives are more specific and tangible than the general goals of a course or even the goals of an individual class. The goal of a class might be, for example, to understand the cell cycle. Corresponding objectives would be for students to be able to define its various phases, to understand the relationship of one phase to another, and to relate this knowledge to disease conditions characterized by abnormal cell division.

It is crucial that your lecture add significantly to what students would get from the assigned readings or the class notes. Otherwise, you can scarcely blame those who do not attend. One obvious way to add value is to update the material in the text, particularly in rapidly advancing fields. Another is to incorporate information from different sources or to connect knowledge in different disciplines. The most common purpose of lecturing, however, is to present material in a clearer and more interesting fashion than the textbook and to highlight the most important points.

A good lecture begins with careful planning. You need to outline your lecture, in terms both of content and of time. Many beginners, and even some long-time veterans, attempt to include too much and end up rushing through their last points or running out of time. Do not prepare detailed notes to be read verbatim; instead, write down the main points you plan to discuss. Reading notes invariably decreases your eye contact with the audience—a big mistake. It also tends to be monotonous and, in a dim classroom, creates a dangerously high zzz factor. Less dependence on lecture notes also allows you the freedom to move around the room.

Do not shy away from using the board to write down main points and simple diagrams that will help students understand connections between different aspects of the lecture. Writing on the board gives students a moment to catch up, and it provides an additional cue for learning that may help them remember the main message. Other cues that facilitate learning include the use of clever examples and metaphors. (A sense of humor is good, but avoid inadvertently insulting somebody or hurting their feelings.) Alter your tone of voice and speed of delivery to emphasize key points. All of this makes the information more memorable and helps keep students energized. We all have had teachers with boring lecture styles. Be sure not to base your teaching manner on these instructors.

It is a good idea to summarize major points and conclusions at the end of each section of the lecture. Mention connections between different segments to smooth the transitions and maintain the logical flow of the presentation. Do not leave this until the end of class. At that point students are busy putting things in their backpacks, checking their cell phone messages, and getting ready to leave for another class. You will be talking to yourself.

You should also periodically check students' learning. Refresh their attention every fifteen to twenty minutes. You might start by inviting questions, but don't be surprised if you have no takers. Often this is simply because students have not had a chance to absorb and digest the material just presented. Others are shy or concerned that their peers might consider their question dumb. One approach that works well is to give students a minute to write down a summary of the preceding lecture segment, and then invite questions. When you do get questions, answer those you can; if you don't know the answer, admit it but promise to find out and get back to the class. Other ways of checking students' learning include projecting one or two multiple-choice questions on the screen and asking for a show of hands in support of each answer. Asking questions and seeking volunteers to answer is less intimidating than pointing to a specific person. Be careful never to humiliate a student for giving the wrong answer. What you are gauging here is how successful you have been in delivering a clear message. Follow up by clarifying any misunderstandings and ambiguities. And be on the watch for body language—yawning, fidgeting, rolling eyes—that indicates confusion or boredom. Taken together, these approaches differentiate merely delivering information from bringing about real learning. Make learning your goal.

The end of a lecture is as important as its beginning. Summarize the take-home points and put them within the context of the general objectives of the course. Propose a couple of issues for students to contemplate in preparation for the next lecture. A tried and true method to get students to listen carefully to this wrap-up is to introduce your points as "what I would be thinking about when constructing exam questions related to today's lecture." At the end of class stick around to answer questions. Some students prefer coming to the podium to ask a question rather than asking it in public. Others save their questions until the end of the lecture in case it is answered along the way.

Technology and teaching

Modern technology provides great tools for teaching, starting with the wireless microphone that allows you to move around and keep the presentation active. Your laser pointer is handy for highlighting important content on the screen. Advanced audiovisual technology makes it possible to show movies of real laboratory experiments or fieldwork. Using the Internet, you can show relevant Web sites and navigate through them at amazing speed. Most importantly, you can teach students at multiple sites simultaneously and in an interactive fashion.

The advent of electronic tools for use in presenting lectures, notably PowerPoint, has made a teacher's life much easier. Using this technology enables you to show your outline on the screen as a reminder of what you plan to talk about. It also allows you to update your lecture painlessly. You can even use it to show questions to test students' learning at the end of each major section of the lecture. Be aware, however, of the drawbacks of such presentations. The most serious pitfall is the natural tendency of both presenter and students to focus on what's shown on the screen. This creates monotony that reaches its worst when the lecturer is simply reading off the screen. A good solution is to include only a list of bullet points to use as a guide in your delivery, combined with slides of information analysis and the main take-home message.

The following are useful online guides to electronic presentations in the classroom:

- Alice Christie, "Using PowerPoint in the Classroom," http://www .west.asu.edu/achristie/powerpoint/
- Claremont McKenna College Teaching Resource Center, "Evaluating Student PowerPoint Presentations," http://www.cgu .edu/pages/762.asp
- Frances Condron, "Using PowerPoint in Teaching," http://www.oucs .ox.ac.uk/ltg/reports/ppt.xml
- Connecticut College Center for Teaching and Learning, "Teaching and Learning with PowerPoint," http://ctl.conncoll.edu/pp/
- Michael Russell and Walter Shriner, "Creating Effective PowerPoint Presentations," http://www1.umn.edu/ohr/teachlearn/tutorials/ powerpoint/resources.html
- Robert Sommer, "Projector Blues" (*Technology Source*, November/ December 2003), http://technologysource.org/article/projector_blues/

Lecture handouts

Preparing lecture notes to distribute to students at the beginning of a lecture is common practice. It allows students to take notes on key concepts effectively and efficiently, rather than racing to transcribe basic facts and details. It also allows you to move faster through lecture content. It is not a good idea, however, to make these notes so detailed that reading them could substitute for attending the lecture. Taking notes is an important cue for learning and gives students a chance to summarize information in their own language, consistent with their own level of understanding. It is better, therefore, to include major headings followed by ample spaces for taking notes. Including unannotated graphs and diagrams to which students can add notations will also save precious time.

To lecture or not to lecture: Active learning

Formal lecturing remains the most popular and widely accepted method of teaching in higher education. However, this method provides limited types of input to brain centers involved in learning and memory. Neuroscientists tell us that one learns and remembers less when passively listening to a speaker. The brain remembers not only content, but also the process of listening to spoken words and writing them down. A dynamic, charismatic teacher who varies the pace of the presentation and uses interesting examples to highlight important points offers additional input to the brain, which can potentiate the signal triggered by listening. Nonetheless, formal lecturing usually entails a one-way delivery of information, which results in a type of learning low on the scale of Bloom's taxonomy.

Modern trends in teaching, therefore, urge that students actively participate in their own learning experience. Discussion is one of the most effective tools of this active-learning approach. It gives students an opportunity to think of the information that is being presented and analyze it to arrive at take-home conclusions. It also familiarizes students with basic logic, concepts, and principles and with ways they can be applied to solve problems. Through discussion, students evaluate the validity of information and weigh it against opposing evidence. This approach provides varied modes of input to neuronal circuits involved in learning, which help fix information in the memory bank. In technical terms, discussion moves learned information from short-term to long-term memory by connecting it to existing frameworks of knowledge. Analysis and evaluation are, as we have noted, higher levels of learning than the mere accumulation of knowledge. An added benefit is that discussion gives teachers a clearer picture of the level of students' understanding of the material.

Conducting an effective class discussion requires clever, well-crafted strategies. First, establish ground rules and expectations and set clear goals. Then pay attention to discussion dynamics. Some students might be shy or easily intimidated by others. To get everyone involved you will need to engage these students and prevent more outgoing or attention-hungry students from dominating the discussion. A case study or a set of questions about a current controversy in the state of knowledge can provide a fruitful starting point, encouraging problem-based learning that relates pieces of basic knowledge to issues that students may encounter on the job. This in turn promotes interest in learning and understanding the material. Some students are more able than others to come up with and articulate points in the course of a spontaneous discussion. It helps to give students a couple of minutes to write down their thoughts. With experience you will develop skills in kindling discussion and keeping it going without straying too far from the main goals. The trick is to continue to pose more challenging questions and to contrast points of view. Always remember to integrate discussion within class structure and to time it so that you are able to meet your benchmarks for each class. It is also important to summarize the results of each discussion and how they relate to course objectives.

There are many variations on the theme of using discussion for active learning. You could divide a large class into discussion groups, assigning each group a different problem or a different task related to the same problem, then have a representative of each group present its main conclusions to the rest of the students. This approach results in valuable learning from peers. Some teachers use debate between groups of students to develop understanding of important concepts. Be creative. Apply your own novel approaches to achieve active learning.

Testing

Testing is one of many means of assessing students' learning, and it should be structured to serve this goal. The specific goals and objectives of your course should guide you in deciding what type of questions to construct. Multiple-choice questions, for example, assess simple knowledge of facts. To assess deeper levels of learning they must be supplemented with other types of questions, for example short-answer items, problems, and essay questions. Testing can also provide an additional means of learning. You might, for example, consider open-book or take-home exams, which allow students to use all available resources to answer a question and encourage them to discover and assimilate new knowledge. These types of tests mimic real-life situations, where one is allowed to consult the literature and other resources rather than relying wholly on memorized facts. Bearing this in mind, you might go even further and allow students to work in groups on take-home exams. This, however, requires that you design ways to ensure that all group members' contributions are reflected in the answer.

Plan several tests scattered throughout the course, preferably one at the end of each major section. Consider whether to assign equal weights to all tests or to give a lower weight (or none) to the first exam or that with the worst score. Testing pays its highest dividends when you share with students your philosophy of testing and grading as a means of assessing learning.

Welcoming diversity in the classroom

Students in higher education come from a variety of racial, social, economic, and educational backgrounds. This creates a rich environment for

shared learning; to maximize the benefit, make sure your class environment is inclusive and inviting to all. Mix students of different backgrounds
when assigning discussion groups. Do your best to accommodate the special needs of students with any kind of disability. Insist on respect for the
opinions of others, and never allow insulting or stereotypic comments by
any student.

Students also differ in their preferred mode of learning. Some learn
better from pictures and diagrams, others from written or spoken words.
Some focus on facts, others on concepts. Some prefer to study in groups,
some to study alone. Your teaching style has to cater to all of these varied needs. This is best accomplished by presenting the same information
in different ways, for example, by showing an outline followed by verbal
comments and writing summary points on the board. A clever and experienced teacher does this in a smooth way that avoids the appearance of
redundancy.

Students with problems, and problem students

Teaching can pose challenges that go beyond how to effectively convey knowledge and foster learning. Students may come to you with complaints—
some valid, some not—about other students or even other teachers. Listen
carefully and respectfully to their complaints, but keep in mind that one
side of a story often diverges from the full reality. Also, be sensitive to the
fact that the complaining student might be concerned about confidentiality
and worried about the potential for retaliation, especially if the complaint is
related to your own teaching style, accent, grading, etc. You should let the
student realize that you appreciate his or her candidness since it enables you
to become a better teacher. If a student complains about another teacher,
provide advice but avoid disparaging your colleagues. In such a situation, it
might be best to refer the student to an ombudsman or dispute resolution
service.

Remember that it is your responsibility to maintain a positive learning
environment in the classroom. It is most effective to prevent problems
before they arise. Clearly spell out in the syllabus your expectations for classroom behavior. Students will appreciate a well-run, well-organized classroom and the chance to participate in active learning. If you find, however,
that a student is dominating the discussion or is inappropriately aggressive
or disrespectful to you or others, it is best to address the situation imme-

diately. In most cases, you should talk to the student privately after class to let him or her know that such behavior is not conducive to a healthy learning environment and will not be tolerated. Restate your expectations for appropriate classroom behavior. However, be cognizant of cultural differences in communication style (e.g., hand gestures, high voice) before you judge a student's behavior as aggressive or disrespectful. In rare instances, you may feel that a situation is unsafe. In such a case, you should ask the student to leave the classroom, calling university security, if necessary, for assistance.

Apprise your students of proper academic conduct and of the penalties for violations of the university's student conduct code. If you suspect a student of such a violation, consult with your supervising faculty member for advice and assistance. Intentional cheating could be grounds for failing a class or for dismissal. However, students who cheat sometimes do not intend to do so, but are careless or confused about the standards.

You may occasionally find students who are inattentive, unprepared, or discouraged. These students also deserve your special attention. Meet with them individually to discover the basis of their problems, and propose ways in which they might become more engaged with the course material and improve their performance. Suggest a tutoring service if you believe it would be helpful. Be aware that there could be a psychological basis to their problems, for example, depression or a learning disability. You may wish to inform them of university counseling and consulting services, where a well-trained professional staff can help students assess the scope and nature of their concerns, provide learning and academic skills assistance, and give individual counseling for more significant provisional concerns. Note, however, that you must not tell a student to "get counseling" or presume to diagnose a learning, physical, or psychological disability. The decision to seek counseling and the corrective actions to be taken should be made by the student in consultation with counseling professionals. Students identified as having learning or psychological disabilities may register with disability services at their university and receive appropriate accommodations. Helpful guidelines can be found at http://www.mentalhealth.umn. edu/facstaff/general.html.

Two students at our university have prepared a useful paper (Gawne-Mark and Miller) to help graduate instructors solve a range of classroom problems.

Team teaching

Your teaching portfolio might include teaching part of a course taught by multiple instructors. Your contribution may vary from a couple of lectures to a significant portion of the course, depending on how many instructors are involved. Courses designed for graduate and professional students often have many instructors led by a course director or coordinator. Such distribution of effort requires special preparation on your part. You must familiarize yourself with the course syllabus, particularly its sequence of topics. Contact the course director and other instructors to discuss the content of other lectures and potential overlap. You might even find it beneficial to attend a couple other lectures, particularly those immediately preceding yours. This will help you connect your presentation to the others and will therefore enhance student learning.

Teaching a laboratory

Laboratories have always been an integral part of science discovery and teaching. Their premise is straightforward. In one format you as an instructor perform a demonstration; in another the students do an experiment. In either case, discussion of the results follows. As easy as it might seem, running an effective and informative laboratory session requires careful thought and preparation. Without this, your students may regard their lab work as a cookbook exercise and possibly a waste of time.

Make sure all required reagents are prepared and equipment set up and calibrated to save time and make the lab interesting. As in planning a lecture, prepare a succinct and informative introduction to the lab. Remember, this is not a lecture, so be brief. Outline the purpose of the demonstration or experiment and how it ties in to the overall course. Walk students through the steps of the experiment. Mention the timeline of various segments of the lab session. Remind students of safety precautions. Lab sessions can be disastrous if things go awry. Results might not turn out as expected. Accidents might happen. Plan how to respond.

Demonstrations can be valuable tools for proving concepts. Think carefully about what type of demonstration will be most effective in achieving this goal and most suitable given the class size, level of the course, and available resources. What preparatory readings should you assign to students? Would showing a video of the demonstration suffice? How can you keep students engaged and motivated to learn while watching the demonstration? What questions should you ask and when should you ask them in order to test student comprehension? What will you do if things do not go as well as planned?

Never leave the lab while students are performing an experiment. Never bring reading material or exams to grade. Instead, you should walk around and interact with students. Find out how far they are doing and what results they have obtained. Answer questions. More importantly, ask questions to test their understanding of major concepts. Look for signs of confusion and intervene to help.

Always follow a lab demonstration or experiment with a discussion of the results and observations. How do students interpret the outcome? How do they explain differences in experimental outcome between different groups of students? How did the lab exercise deepen their understanding of underlying theoretical concepts? How could they use the same experimental setting to test related concepts? Plan an effective closure to summarize what was learned during the lab session.

Take-home messages
- Types of learning range from gaining knowledge to being able to analyze, evaluate, and synthesize knowledge.
- Presenting material in more than one way will enable you to reach students with varied backgrounds and learning styles.
- A carefully designed course syllabus will minimize obstacles to effective teaching.
- Develop good lecturing habits, and engage students in discussion to promote active learning.
- Use teaching technology to enhance learning, not as a gimmick.
- Design tests with course goals and objectives in mind.

References and resources
"Designing Smart Lectures," http://www1.umn.edu/ohr/teachlearn/tutorials/lectures/.

GawneMark, Chelsea, and Ethan Miller. "Student Conduct Issues: Action Guidelines and University Resource Flowchart for the Graduate Instructor," http://www1.umn.edu/ohr/img/assets/23652/conduct.pdf.

"Major Categories in the Taxonomy of Educational Objectives" (summary of Bloom's taxonomy), http://faculty.washington.edu/krumme/guides/bloom1.html.

National Research Council. 1999. *How People Learn: Brain, Mind, Experience and School.* Washington, DC: National Academy Press.

National Research Council. 2001. *Knowing What Students Know: The Science and Design of Educational Assessment.* Washington, DC: National Academy Press.

6 DESIGNING YOUR POSTDOCTORAL EXPERIENCE

What is postdoctoral training?

A postdoctoral appointment, as defined by the Association of American Universities, provides transitional training in the period following completion of a PhD or other doctorate (e.g., ScD, MD). The appointment is temporary, not part of a clinical training program, and intended as preparation for an independent academic or research career. It is a critical period of apprenticeship. Postdocs work more or less full-time on research or scholarship, under the supervision of a senior scholar or department at a university or other research institution, and are free (indeed, expected) to publish the results of their work (http://www.aau.edu/reports/PostdocRpt.html).

A doctorate is the highest degree awarded by universities. Why, then, would you spend more time in training? Why not move directly into a good job that pays better for fewer hours of work? A simple answer is that employment in some disciplines requires postdoctoral experience.

Taking a postdoctoral appointment has some notable advantages. As a postdoc you will have more independence and command over your research project than you did as an undergraduate. You will learn new research skills and enhance your curriculum vitae or résumé with more publications. If you plan your postdoc training wisely, you can gain many transferable skills, for example, how to teach, write grant proposals, and manage a laboratory group. This will let you hit the ground running in the next stage of your career.

There are also some disadvantages to postdoctoral training, not the least of which is financial. While there has been a general trend toward higher postdoc stipends, they are still markedly lower than average salaries of jobs in science and engineering that require only a bachelor's degree. Furthermore, the stipend gap between graduate students and starting postdocs has been narrowing steadily, making the financial jump many students anticipate upon graduation less pronounced. Of course, low pay has a particularly noticeable impact on postdocs who have family financial obligations. The

National Postdoctoral Association estimates that 69 percent of postdocs are married or otherwise partnered, with 34 percent having children. Postdocs are expected or driven to put in long hours in the lab, which may leave little time for personal or family life. Furthermore, the status of postdocs within academic institutions is often nebulous. They are neither students nor faculty, and this can result in a dearth of institutional services and less independence in directing their research projects.

Should you pursue postdoctoral training?

Whether you must do postdoctoral training depends on common practice in your specific field and the type of job you are seeking. For instance, someone with a recent PhD in many fields of engineering or in pharmaceutical sciences can usually get a faculty position or a research job in an industrial firm without postdoctoral training. If you are in one of these disciplines, we would not recommend taking a postdoc unless you feel the need to expand your research credentials. If your PhD is in the life or physical sciences, however, you will most likely have to complete a postdoc before you can compete for high-level jobs. According to the National Academy of Sciences, approximately 60 percent of new doctorates in the biological sciences and 45 percent in physics, chemistry, or astronomy plan to obtain postdoctoral experience. (This is in sharp contrast to roughly 15 percent in engineering.) Browse job advertisements to get a sense of the importance of postdoctoral training for various types of employment in your field.

The number of years spent in postdoctoral training also varies by discipline. Postdocs in biological sciences, physics, and astronomy spend an average of four years in postdoctoral training, those in engineering one and a half years, and those in chemistry two years. Some people complete more than one postdoc stint; according to the National Postdoctoral Association, the most common reasons for doing so are to gain additional research experience or to work with a specific advisor, followed by a lack of other employment opportunities (http://www.nationalpostdoc.org).

Not all lines of employment for life science PhDs require postdoctoral experience. You may land a job, for example, as a representative for a pharmaceutical firm or a teacher in a small college without completing a postdoc. Even if these directions match your career ambitions, however, you should seriously contemplate the pros and cons of bypassing postdoctoral training. On the one hand, you will be making a decent salary and climbing the job ranks right out of graduate school. On the other hand, you may find you do not like it after a few years; then, in order to change careers, you might have to do postdoctoral training after all. We advise that you pursue a postdoc, even a short one, if the majority of careers in your field require one. The decision is ultimately yours—and must take into consideration, in addition to career aspirations, such factors as age and family needs—but the experience will afford you both flexibility and a competitive edge.

What should you get out of postdoctoral training?

Some envision postdoctoral training as a holding pattern while they wait to get a "real" job. This negative characterization has gained currency as the duration of postdoctoral training has steadily increased in response to a scarcity of jobs. However, postdoctoral experience has many merits. It affords you a chance to become a better-rounded researcher and, if the training, however brief, is carefully crafted and productive, to become more competitive in the job market.

In designing your postdoctoral training you should give a great deal of thought to your career goals and what tools you will need to achieve them. In general, postdoctoral training should prepare you to become a marketable independent researcher. This requires learning new research methods and technology and becoming more adept in experimental design and data analysis and interpretation. In this respect, postdoctoral training should function synergistically with your graduate training. You will also have to publish your completed postdoctoral work. A large number of publications in premier journals demonstrates that your peers in the scientific community value your research and is perhaps the best single predictor of competitive success in the job market, regardless of the type of career you pursue.

You should take full advantage of the opportunities your postdoctoral training gives you to do research without having to take or teach classes and to read more of the literature and keep up with advances in your field. A postdoc also affords more chances to communicate about your research with other scientists, both within and outside your institution. Such communication extends your network of colleagues and may lead to new research collaborations. These networks are invaluable, providing access to materials and ideas and improving your chances of getting a future job. Other transferable skills you may derive from postdoctoral training include teaching, grant writing, laboratory management, and public speaking.

How can you find about available postdoctoral positions?

Postdocs are in demand in some disciplines but not in others. More importantly, good postdoc mentors are much sought after. Thus, you must start the search early, as it requires much soul-searching to identify your specific career goals and find a good match in an advisor and institution.

Some postdoctoral opportunities are advertised in journals such as *Science* and *Nature*, and Web sites like http://www.postdocnet.com can be useful, but the best and most common way of finding good opportunities is through personal contacts. About a year before you expect to complete your doctorate, discuss your career goals with your advisor and other faculty members and ask them to suggest potential postdoctoral advisors. Network with senior researchers at scientific meetings to find out if they have or anticipate openings in their labs in the near future. If you are targeting a highly prestigious lab, you should expect to be required to apply for an external fellowship to support your training. This requires additional preparation. Needless to say, such labs are in great demand and fill up quickly. Participate in job fairs, which are often organized by professional societies during their annual meetings. This will give you wide exposure to a range of opportunities and a chance to meet potential advisors face to face. Consider these exchanges mini-interviews that could lead to interesting future possibilities.

Where to postdoc?

Success in postdoctoral training is key to all future steps in your career, so you should choose your institution and advisor with the utmost care. It may be tempting to pursue postdoctoral training in the same institution where you obtained your doctorate. Don't do it! Seek a new research environment where you will be exposed to fresh ideas, approaches, and institutional culture. Avoid the possibility of continuing to be treated as a graduate student. If you absolutely have no other choice (for example, due to family obligations), we strongly advise you to choose a postdoctoral advisor in a different college of your university, or at least in a different department. Your postdoc lab should provide you with distinctly new research skills and approaches. Do not pick the easy route of pursuing a project very similar to what you did for your doctoral thesis.

Before deciding on a specific institution, you need to decide what type of postdoctoral experience you would like. While most go to universities, you might prefer a government lab (e.g., the National Institutes of Health or Los Alamos), an industrial lab, or a research institute such as the Cold Spring Harbor Laboratory or the Institute for Advanced Study. Postdocs in industry and government usually earn more than those in academia, and they often have better resources and a more collaborative environment. As a postdoc in academia, on the other hand, you have an opportunity to gain breadth and experiences beyond doing research, such as teaching and grant writing. Do not limit your options to labs in your own country; seek knowledge abroad if this is what is best for your career.

ACADEMIC, FEDERAL, OR PRIVATE RESEARCH INSTITUTIONS

There are many factors to consider if you choose to do a postdoc in an academic institution. Many deal directly with research and the quality of training: What is the institution's research reputation and national ranking? What resources does it have that are pertinent to your training (e.g., core equipment facilities, libraries, machine shops), and are they of appropriate quality? Is the school surrounded by other, well-funded research facilities? How committed is it to high-quality postdoctoral training? Are there enough postdocs to constitute a postdoc community? Is there a postdoctoral association or an office that deals with postdoc issues and policies? Are there career development programs to teach postdocs how to teach, write grant proposals, and prepare for getting future jobs?

Other questions relate to working and living conditions for you and your family: Does the institution classify postdocs as students or as academic personnel? What health or retirement benefits are there for postdocs? Is there child support? What about spouse relocation support programs? Is the institution located in a city that is safe and attractive, with resources that match your family's needs and lifestyle?

A postdoc in a federal or private research institute is generally similar to one in an academic institution. A possible advantage is that intramural research funds are more secure than external grant funding. Also, some private institutes are very prestigious and have very prominent researchers.

INDUSTRY

Advantages of postdoctoral training in industry include becoming familiar with the process of developing and commercializing discoveries and the opportunity to work as part of a multidisciplinary team. Research in large pharmaceutical companies, for example, requires close interaction between biologists, chemists, engineers, and policy experts. An industry postdoc may also get your foot in the door if you are thinking of pursuing an industrial research career. Be aware, however, that not all firms are inclined to hire from within; some, in fact, have firm policies against doing so.

A major drawback of an industrial postdoc is that you may lack access to information pertinent to your project, some of which may be considered proprietary. You may also have less control over what and when to publish. In some cases, these drawbacks can be avoided by designing a research project that does not involve the company's proprietary interests. However, this might not be possible at a small technology company whose primary goal is to develop a specific software package, biological assay, or other commercial product. Most company resources will then be directed toward this goal, leaving no resources for testing other research questions, however interesting. Doing a postdoc in industry comes with additional risks. For instance, management might suddenly decide to close your department and lay off your advisor. This often happens as a result of mergers, a particular risk if you work for a small company.

OVERSEAS

You might be attracted by postdoctoral training overseas. There are top laboratories in all disciplines of science and engineering in many European countries, Australia and New Zealand offer excellent postdoc opportunities, and Asia is increasingly becoming a place to consider: Japan has been a leader in research for many years, and Taiwan, Hong Kong, and Singapore are developing rapidly. An overseas postdoc provides an opportunity to get exposure to a different research environment and approaches, as well as a different culture. On the down side, research funding in Europe is not as generous as in the United States. The new Asian economies are, however, investing heavily in scientific research. Working overseas will also make it more difficult to attend the many international conferences held in the U.S. This may make it more difficult to keep up with advances in your field, decrease your personal contact with potential future employers, and discourage U.S. employers from inviting you for job interviews due to travel costs. However, modern electronic means of communication—from e-mail to video conferencing—make this less of an obstacle. Overall, the experience could be well worth it, and we know of many who have gone through overseas postdocs with no regrets.

Senior or junior postdoc mentor?

As when choosing a graduate thesis advisor, as discussed in Chapter 3, there are advantages and disadvantages to selecting either a senior or a junior faculty member as the advisor for your postdoctoral training. Senior scientists (full professors or established associate professors) are more experienced and better established in their fields. Their national stature and visibility will facilitate networking with others to establish research collaborations and obtain hard-to-get experimental reagents. You may also gain an edge in competing for jobs. People who do not immediately recall your name may remember you as the postdoc in Professor X's elite group. The name of your advisor (and institution) will always stand behind you; indeed, some believe that fame is often inherited. Many employers will be eager to add blue blood to their workforce.

However, belonging to such a prestigious group usually comes with a cost. You can expect to be one of a dozen or more postdocs and graduate students in the lab, along with a handful of technicians and undergraduate trainees. Your advisor will be endlessly busy writing grants, finalizing manuscripts for publication, and traveling to give invited research seminars or sit on grant review panels or advisory boards. This leaves little, if any, time for one-on-one interactions with members of the research team. In some large research groups, only one or two senior members interact directly with the boss, and directions may be passed down through several intermediate levels. This is not all that bad, since the intermediaries are likely to be highly qualified and take their responsibilities seriously.

In such an environment you can expect to be trained by many people, not just a single advisor. (Indeed, few senior scientists continue to work in the lab or even put on a lab coat except for staged photos, so your advisor may be unfamiliar with details of experimental methodology and unable to help troubleshoot your experiments.) And you will likely have the opportunity to train others. You may have substantial freedom to steer your project, and the availability of generous funding may allow you to pursue new directions and try out risky approaches that could pay big dividends. Your advisor's popularity on the lecture circuit may also lead to your being recommended to give a seminar or two when his or her calendar is booked, affording you much greater recognition than you could achieve on your own.

In such a big group, however, you cannot expect to have a project that is all your own. Most of your work will be done in collaboration with other members of the research team. This will lead to more, and more varied, publications and will teach you new research skills. However, you will have to make certain that such collaborations are organized so that you get a fair share of first-author publications.

By contrast, junior faculty members (assistant professors) will likely be in the lab on a daily basis doing experiments side by side with you. This continual personal interaction and hands-on experience in problem solving can be beneficial, but it can also prove too close for comfort. Excessively close supervision or daily questions about how a particular experiment fits with the goals of the lab (and its funders) might make it difficult for you to carve out your own niche in the research project, something you can carry with you when you start your own lab.

At this early stage in their careers, junior faculty members are often dependent on departmental start-up funds and perhaps one major external grant. Thus, they may be hesitant to "waste" precious resources on risky experiments, however great the potential gains. You might have to do most or all of the work yourself, maintaining apparatus or synthesizing reagents from scratch, with little or no technical help. And you might not be able to attend and present at multiple scientific meetings every year. Such restrictions may slow your learning and reduce your productivity.

The hard work, however, could bear rewards when you have to run your own lab on a small budget and must pursue a similarly conservative approach. Moreover, this tightly structured style will teach you more about the systematic pursuit of a scientific question, which is important in writing successful grant proposals. It will also ensure that you can follow each project to completion and add the resulting publication to your curriculum vitae. Junior advisors are hungry and motivated to publish, always aware of the tenure decision awaiting them in a few years. Most of your publications will probably have only your and your advisor's names. This carries more weight than being one of a dozen authors, even if you are listed as first author.

Which offer to accept?

In selecting a postdoctoral advisor, history is usually a reliable guide, and the degree to which current and past postdocs are satisfied with the training they have received is a good indicator of that history. Before committing to a postdoctoral appointment, you will have to visit the laboratory and have extensive discussions with your potential advisor and labmates. Request ahead of time to meet with as many of them as possible, particularly post-docs. Explore the breadth of research projects across the lab and possibilities for collaboration. Politely probe into their general impressions of the lab environment, but realize that some might feel inhibited to talk frankly in the presence of others. Check out the number and quality of the postdocs' publications, the jobs they have landed, and how long it took them to find their jobs. Ask about the advisor's personal training style, how much guidance and freedom the postdocs were given in pursuing their research projects, and whether the advisor encourages postdocs to seek career-enhancing skills and consider nonacademic careers. Read between the lines, and ask if you could chat with individuals later by phone.

If you do not have the following information before visiting the laboratory, ask for it when you are there:

- A copy of the advisor's curriculum vitae. Check productivity, both recent and past; gaps in publications or external research support should raise a red flag. Don't simply count the number of publications; look for publications in high-quality, high-impact journals. Give more weight to invited presentations at other institutions and at conference plenary sessions than to other conference presentations, which are often not peer reviewed. Note whether he or she serves on the editorial boards of key journals or on national grant review panels.
- Information on the research background of lab personnel, including other postdocs, graduate students, and technicians. A lab in which others share your research interests will provide a support network and opportunities for collaboration. At the same time, however, you want to be able both to learn new research skills and to add to those already present in the lab.
- History of job placement of postdoctoral trainees for the past five to ten years, including names and contact information of past postdocs.
- A list of chores you would be expected to perform in the lab. Is technical assistance provided, or will you be expected to prepare reagents, troubleshoot equipment problems, or maintain cell cultures?
- Lab policy regarding authorship, ownership of ideas, and taking some aspect of your project to another lab when you leave. This must be clearly delineated from the outset. Never take for granted that your

advisor will allow you to pursue the same, or even a closely related, project in your own lab, even if the project is your brainchild.

- Advisor's philosophy regarding the balance between individual and collaborative work.
- Advisor's willingness to mentor you in grant writing and reviewing manuscripts for journals, and to allow you time to build your career.
- The likelihood that you will supervise students and junior scientists in the lab.
- The sources of the funds that support the position. This may determine how much freedom you have in conducting your research. Federal grants, for example, require that you stick to the general goals outlined in the application for funding. Money from individual research fellowships may have fewer restrictions, but can you get help in writing the proposal? If the fellowship is not funded, will your position be in jeopardy? Will funding through an individual fellowship mean that you don't receive institutional employee benefits? Will the stipend be less than you would be paid on your advisor's grant?
- Stability of funding. Labs supported by multiple research grants are better able to continue working if one grant dries up.
- Duration of appointment. If the appointment must be renewed on an annual basis, be clear about renewal requirements. Think twice before accepting a postdoc position in a lab supported by a single grant that is in its final year.
- Availability of funding to attend and present at national and international meetings. Expect to have support to attend at least one meeting every year.
- Starting salary range and the basis of annual merit raises. Do not overlook regional differences in cost of living.
- Benefits, particularly health insurance, for you and your dependents. Also inquire about sick and family leave. You might be referred to a personnel specialist in the department to address these issues.
- Expectations regarding work hours and vacation. (Ask, but don't give the impression that you'll only work a 9-to-5 schedule or that you plan to go on a long vacation soon after you arrive.)

Then go home and digest this formidable amount of information. Take both the institution and the lab into consideration, and compare them to other opportunities you might have. Design a grading rubric that lists all relevant factors. Discuss your options with your doctoral advisor, other faculty members, and peers. Ask for their advice and what they know about the reputation of the institution and the lab. You will likely come to the conclusion that there is no perfect choice, but you can aim at selecting the best opportunity. If you are offered the postdoc position of your choice, make sure the offer letter contains as much specific information as possible about the terms and conditions of appointment, your advisor's expectations, and the basis of evaluation of your progress. You can find examples for postdoc position offer letters and some guidelines at http://www.graddiv.ucr.edu/Postdocs/SampleApptLetterPI.pdf, http://www.nap.edu/openbook/0309069963/html/46.html, and http://rgs.rice.edu/Research/PostDoc/PolicyPostdocs03222000.cfm.

Which research project?

Choosing a postdoctoral research project is naturally tightly linked to selecting an advisor. Your main question should be whether it will be in the same area as the research you conducted for your PhD thesis. Staying within the same general area will allow you to establish expertise and visibility. This is wise—but only if you still like it.

You should not, however, join a lab that asks closely related questions or uses similar research technology. Postdoctoral training is a golden opportunity to expand your research portfolio, particularly in learning new research techniques. If your doctoral research dealt with the behavioral manifestations of street drugs in experimental animals, for example, you might want to expand on it by exploring the signal transduction pathways or genetic elements involved. In contrast, studying the abuse potential of cocaine versus that of morphine using the same methodology would not represent anything new. Resist the natural temptation to limit yourself to what's comfortably familiar.

Consider research techniques and methodologies as the building blocks of your future independent research program. The more techniques you command, the more flexibility you will have in exploring research questions in depth and from many angles. There is nothing more boring than randomly testing the effects of the "drug of the day" on rats' ability to find their way through a maze, or the influence of one substance after another on the melting temperature of asphalt. Interesting research is often, but not always, hypothesis driven and aims at exploring the mechanisms that underlie a given biological or physical phenomenon. This necessitates the use of different experimental approaches.

A second postdoc?

Suppose you are not satisfied with your first postdoctoral experience, either because of inadequate productivity or because it is not leading to a satisfactory job. You must now make a major career decision. First and foremost, should you stay in research or try something different? If you decide to stay the course, how long should you remain in your first postdoc appointment before seeking another?

Many academic institutions have policies that limit the duration of a single postdoctoral appointment to three to five years. This is mainly to protect postdocs against abuse by advisors who might want to keep experienced postdocs much longer than is appropriate, taking advantage of their talent and productivity. These policies also remind postdocs that they are there for training; a postdoc is not a job per se but an apprenticeship intended gradually to transform them into independent researchers. We advise postdocs who plan on another stint of training not to spend longer than three years in their first position. This is long enough to judge whether they are ready for the next stage of their career.

If you decide to take a second postdoc, what type should you pursue? Do you want to change from academia to industry or government, or the other way around? Depending on your career ambitions, such combinations of experience might have a synergistic effect. In any case, the goal of learning something new still applies.

While a second short postdoc might be necessary to become competitive in a tight job market, we strongly advise against doing a third. Generally speaking, the total duration of your postdoctoral training should not exceed five years. Any longer might be harmful to your career.

International postdocs coming to the United States

If you are coming to the United States from another country to pursue a postdoctoral position, you will face additional challenges: different work habits and social customs, possibly a language barrier, and even different definitions of postdoctoral research. Difficulties usually arise early in the process of applying for postdocs from afar. You will not have the opportunity for face-to-face exploratory conversations with potential mentors, unless you meet at an international conference. If you are about to obtain a doctorate in your home country and desire a postdoc in the United States, make the fullest possible use of such precious encounters. Dedicate a significant portion of your time at the conference to networking with potential advisors. Give them copies of your curriculum vitae and publications. Take similar advantage of any visits by U.S. scientists to your institution, following up with e-mail to reiterate your interest in a postdoc position and inquire about openings.

Your American host institution will provide documentation of the fact that you have been offered a postdoctoral position, specifying the terms and conditions of the appointment, and will help you get a visa to enter and work in the United States. Familiarize yourself with the differences between various types of visas, particularly in regard to the length of time you will be allowed to stay and your spouse's eligibility to work. The National Postdoctoral Association has a comprehensive guide to visas and immigration status for international postdocs on its Web site: http://www.nationalpostdoc.org.

If English is not your mother tongue, start working to improve your language skills as soon as you decide to pursue a postdoc in the United States. While you will not be required to take a formal language proficiency exam such as the Test of English as a Foreign Language (TOEFL), perfecting your English, especially your listening and speaking skills, will serve you well both in and out of the lab. Watch American movies or TV programs to get used to American accents. Take note of colloquial expressions and idioms and find out what they mean. Practice speaking, ideally with an American who lives in or is touring your country. When you arrive in the United States you may be relieved to find many others who speak your native language, but do not fall back into your comfort zone. Continue to speak English, even with them. A consistent observation in our own research groups is that foreign postdocs who room with U.S. students or postdocs acquire language skills much faster than those who live with others from their home country. This might be less of an option if you are married or living with a significant other. In this case you could compensate by socializing with American friends. Watch U.S. television, particularly the news. This will familiarize you not only with recent events, but also with American politics, idioms, and ways of life.

You will learn a lot more once you arrive, and some of this knowledge may surprise you at first. For instance, you will likely be asked to stop calling your advisor "Sir," "Professor," or even "Doctor." Most American scientists prefer to be called by their first names by their students and postdocs. When you arrive at your new department, you are not likely to find your professors dressed in three-piece suits and ties. Get used to blue jeans and plaid shirts, even shorts and T-shirts in warmer climates. There are exceptions, of course, but most dress and interaction will be informal. Do not wait for the "boss" to talk to you and ask about your progress. Most advisors will have an open door policy unless they are particularly busy preparing for a committee meeting, giving a lecture, or writing a grant proposal. If they are busy, they will let you know. You should also try to schedule periodic meetings with your advisor to maintain consistency and efficient communication.

There are some major differences in how postdoctoral research is defined in the United States and in many other countries. Most American postdocs are trainees who have just obtained their doctorates. Many foreign postdocs in the U.S., particularly those from Asian and African nations, are, in contrast, midcareer and senior foreign scientists seeking to add to their research skills or to beef up their publication records in order to gain a promotion at their home institution. Some might be alarmed or disappointed by the informal style of communication in an American laboratory, particularly the lack of concern for titles and ranks in conversation. This might be particularly noticeable in regard to shared lab chores where everybody is expected to clean up their own lab space and to take turns filling the distilled water jug or maintaining shared equipment. In some big labs a senior technician is given authority to streamline lab operations, order experimental reagents, or schedule time on heavily used equipment. A full professor from a foreign country might be offended and reluctant to follow "orders" from a person of lower rank. Get used to it.

Negotiate salary with your advisor before you accept the appointment. Some foreigners are shy to talk about financial matters out of respect and to avoid giving the impression that money is a central goal. This is not the case in the United States. Most labs in the biological sciences use the pay scale set by the National Institutes of Health for individual postdoctoral training fellowships. These rates are updated annually and published online at http://grants.nih.gov/training/nrsa.htm. In the physical sciences, the National Science Foundation and the national laboratories set the scales.

Understand, however, that postdoc salaries are largely regulated by institutional guidelines rather than firm policies, which gives your advisor some latitude in determining how much to pay you. Do not get excited by a salary figure without understanding its purchasing power in the United

States. Communicate with friends in the U.S. to get a sense of what starting postdoc salaries mean in practical terms, particularly in the city where you will be living. You should also discuss with your advisor the possibility of partially or totally covering your travel expenses to the United States.

The office for international scholars in your U.S. host institution can answer your questions about housing, transportation, and other concerns. Maintain regular communication with this office, and stay in touch once you arrive in the United States. It is your best source of vital information on changes in visa conditions and policies. It will also explain your tax obligations, help you obtain a social security card, and arrange health insurance for your family.

If possible, obtain a major credit card (e.g., Visa or MasterCard) as soon as you arrive. Credit cards can be used to pay for almost anything in the United States. More important, some transactions, such as renting a car or reserving a hotel room, *require* a credit card, even if the payment is to be made in cash. Credit cards are less popular in many other countries, in part due to their high annual fees. Also, many American banks are reluctant to offer lines of credit to foreign nationals. Do not despair. There are ways for you to get a credit card. One is to apply for joint membership in the National Postdoctoral Association and the American Association for the Advancement of Science. This qualifies you to apply for the AAA's credit card. You can also apply for a secured credit card; with this type of card expenses are charged to a savings account. These cards are easier to obtain but might bear significant application fees and high annual fees.

When you travel to the United States, bring with you important documents such as medical and immunization records, drug prescriptions, credit history, and driver's license. Within the country, some states allow you to drive using a license from your home country in addition to an international driver's permit. Others require you to apply for a state driver's license within a certain time following your arrival. Again, counselors in your institution's office for international scholars can give you helpful guidance regarding these matters.

Take-home messages
- Think of postdoctoral training as an apprenticeship that will prepare you to become an independent scientist.
- Choose an advisor, project, department, and institution carefully to ensure a successful postdoc experience.

- Consider the relative merits of working with a junior or a senior scientist.
- Inquire about the training history of potential advisors and job placement of past trainees.
- Learn as many new techniques as possible. Do not limit yourself to what you already know how to do.
- Develop both professional and personal skills to prepare you for the type of career you would like to pursue.

References and resources

Althen, Gary. 2003. *American Ways: A Guide for Foreigners in the United States*. 2nd ed. Yarmouth, ME: Intercultural Press.

American Physiological Society. "Your Postdoctoral Training: General Information." http://www.the-aps.org/careers/careers1/Postdoc/postgenl.htm.

Association of American Universities. http://www.aau.edy/reports/PostdocRpt.html.

Burroughs Wellcome Fund and Howard Hughes Medical Institute. 2004. *Making the Right Moves: A Practical Guide to Scientific Management for Postdocs and New Faculty*. Research Triangle Park, NC: Burroughs Wellcome Fund; Chevy Chase, MD: Howard Hughes Medical Institute.

National Academy of Sciences, Committee on Science, Engineering, and Public Policy. 2000. *Enhancing the Postdoctoral Experience for Scientists and Engineers: A Guide for Postdoctoral Scholars, Advisers, Institutions, Funding Organizations, and Disciplinary Societies*. Washington, DC: National Academy Press.

National Postdoctoral Association. "International Postdoc Survival Guide." Accessible via "Publications" link at http://www.nationalpostdoc.org.

7

PREPARING FOR YOUR FIRST REAL JOB

Very few careers place as many demands on your time as being a research scholar. As a result, many postdocs make the mistake of postponing preparation until they actually start looking for a job. By then, in our opinion, it might be too late. Early groundwork is perhaps the most important element of successful job hunting. But finding time to do the necessary work takes initiative. Once you start your postdoc appointment, your job-seeking expedition is only a few years away.

Build up your professional skills

Being a successful independent scientist requires a variety of skills. Different combinations of skills are required for different careers, as will be discussed in more detail in the next two chapters. The following is a brief summary.

Research/scholarly skills. You should have diversified research skills without being a jack of all trades and master of none. As a postdoc you may feel more comfortable bringing the technical knowledge you obtained in graduate school to your mentor's lab than you do seeking additional knowledge from the new lab. It is a natural temptation for postdoc mentors to go along with this arrangement since it adds a new dimension to their research programs. The risk, however, is that you will limit the scope of research questions you are prepared to address. It is important that you expand your horizons during your postdoctoral training by learning new research techniques, from your own labmates and from others in your department. Knowledge of a wide spectrum of experimental techniques is especially important for getting a job in industry. But whatever job you ultimately decide on, mastering a variety of cutting-edge technologies will give you greater freedom to pursue more complex research questions.

Publications. In assessing your productivity, those with the authority to hire you will likely consider both the quality and the quantity of your publications. In a sense, a scientist is only as good as his or her last publication, so it is crucial that each be of a high quality. At the same time, you should avoid putting all of your eggs in one basket—a forty-five-page paper with twenty graphs and ten tables, however excellent, will still only be seen as one paper. Keeping this in mind, be creative in pursuing ways to legitimately build up your publication record. Make sure the people around you know who you are and what special skills you can offer; this may facilitate your making significant contributions to projects managed by other postdocs, students, or technicians within your mentor's laboratory (but don't push in where you're not wanted) or to establishing collaborations with colleagues in other laboratories. This approach also demonstrates that you are a team player, which bears particular importance in industry. Collaborative agreements should, of course, be made only with the blessings of your postdoc mentor and must be consistent with the specific goals of the funding source that supports your project, particularly in terms of your effort distribution. You might also offer to supervise students or technicians in the lab; many busy mentors of postdocs would welcome such an offer if they thought the postdoc had appropriate managerial skills and scientific maturity. And if your contribution to the supervised project is significant, it could earn you authorship credit. Finally, you can actively seek opportunities to help write review articles and book chapters with your postdoc mentor; this kind of publication is a prestigious addition to your curriculum vitae.

Grant applications. You should start planning and writing your first research grant application while still a postdoc. Having a grant in hand will most certainly enhance your marketability. Furthermore, writing a grant proposal will force you to go through the invaluable exercise of moving beyond contemplating different approaches to identifying a specific and focused research goal. Expect to write numerous drafts, each hopefully better than the previous one. Make use of the experience of your postdoc mentor and others in the arena of grantsmanship. Consult with those in your department or college who serve on local and national grant review boards; they could give you valuable insider's tips.

Teaching skills. Ideally, you should demonstrate knowledge and experience in teaching to different audiences (e.g., undergraduate, professional, and graduate students). Take advantage of any courses on teaching offered at your or a neighboring institution. Having served as a teaching assistant while in graduate school provides valuable practical experience. If you did not have the opportunity to do so, we advise you to volunteer to teach while you are a postdoc. For example, your advisor might not mind sharing a series of lectures on a topic within your area of expertise. An added benefit of this arrangement is that it gives your mentor an opportunity to observe your teaching. This will enable him or her to make specific comments regarding your teaching skills when writing letters of reference. Another approach is to volunteer to teach at a community college in your area or to substitute for an instructor at your university or a nearby institution during a sabbatical leave. Again, you must get your mentor's approval before making any promises or formal agreements.

Service and leadership skills. Go the extra mile. Volunteer your services on departmental, college, or university committees. Actively participate in your institution's postdoctoral association. Become a leader. This will give you a unique opportunity to find out what goes on behind the scenes and how certain decisions are made. Interaction with committee members might also lead to getting another strong letter of recommendation describing your leadership aptitude and good citizenship. Be selective, however, and only serve on committees that are of interest to you. Learn early to say no if the service would interfere with the progress of your research.

Supervisory and managerial skills. Many scientists are much better at designing a perfect experiment than at running a happy and productive lab. Managerial experience and people skills are usually not a formal part of the graduate school curriculum or postdoctoral training. It will be to your advantage to find a way to gain such experience, no matter which career direction you later choose. Check the offerings of your institution's human resources office. These offices often make supervisory training available to all employees, a benefit that might ordinarily escape the attention of postdocs. Ask your mentor if you could advise an undergraduate or graduate student during their summer laboratory rotation. Mentors often approach their talented postdocs to ask if they would take on such a role, but some wait for the postdoc to express an interest, Again, clear communication with your mentor is invaluable. Learn from your advisor how to handle research budgets, how to make hiring decisions, and, most importantly, how to resolve conflicts. Do not shy away from asking technical lab personnel about the cost of reagents and equipment and how they go about comparing commercial sources for pricing and quality.

Communication skills. You will need effective communication skills regardless of the kind of job you choose. Your ability to teach, advise students, and write fundable research proposals absolutely depends on it. Clear spoken and written communication is essential for your daily interaction with peers and administrators at all different levels of the hierarchy. You should also be capable of conversing with members of the public to share with them the importance of your scholarly work and the mission of the university; to this end, it is useful to contemplate the "real-life" importance of your research and to attend presentations that address practical applications of findings in your field. Most notably, superb communication skills are crucial in convincing someone to hire you. Always be ready to provide a one-minute elevator talk about your skills and training background. You can never tell when or where a golden opportunity will arise. Jack Griffin's book *How to Say It at Work* is a useful practical manual of verbal communication. (See our chapters 17 and 20 for more on communications and speaking.)

Establish uniqueness. Unusual but relevant skills can help distinguish you from the majority of other job applicants. Having completed a course on intellectual property or a short internship in the university office of patents and technology marketing, for example, would enhance your application for a job in industry relative to those who are technically skilled but do lack a clear understanding of what happens in the end zone. Using unique combinations of technology and methodology in your research can also set you apart. If you are a behavioral pharmacologist, learn how to determine genetic variability. Venture into learning computer-based simulation to complement your knowledge in biophysical measurements. This approach requires flexibility and extra work, but the gain is worth it.

Contemplate your dream job

A PhD offers you many career possibilities, which will be discussed in the next chapter. Early on during your postdoc, find out about as many career choices as you can. Discuss career options with your advisor, department chair, and other faculty. Spend time mulling over what you would like to do in the ten years following your postdoctoral training, and perhaps during the ten after that. Personal and professional circumstances often change over the course of one's career, sometimes in unexpected way. However, careful career planning increases your chances of finding your own path and being in command of such changes. Keep an open mind in contemplating possibilities. Never say never.

Among the possibilities for employment, many of which we've previously mentioned, are teaching in a small college; research and teaching in a large institution; research in a corporate or government lab; regulatory or policy administration; work for funding agencies; intellectual property; technical or popular science writing; print or broadcast journalism; business development, consulting, sales and marketing, or executive recruitment; starting your own business; informal science teaching; and conservation work. In sorting through your options, think of these questions:

- What motivates you?
- What are your strengths and weaknesses?
- What is your dream job? What are your future ambitions? What would you like to brag about at a college reunion ten years from now?
- What are your personal values? Do they support or clash with certain career paths?
- What is your optimal career/life balance?
- What are your geographical limitations, if any?
- What are your financial needs, aspirations, and obligations?

Analyze your strengths and weaknesses

Carefully assess what you excel at and what needs improvement. This will make it easier for you to assess the suitability of different career choices. Be realistic. Be honest with yourself, but not overly critical. It is wise to use a numerical scale for each attribute, since this should not be an all-or-none judgment. The Federation of American Societies for Experimental Biology has constructed a useful assessment instrument that could help you in this regard; its Individual Development Plan for Postdoctoral Fellows is online at http://opa.faseb.org/pdf/idp.pdf.

Understand the skills required for different career paths

You need to tailor your postdoctoral training to enhance skills specific to your desired career. In this chapter we will discuss skills necessary for success in academic and industrial research jobs. Other careers in science will be discussed in chapters 8 and 9. A job in academia, for instance, requires proficiency in teaching at many levels, as well as solid published research. Good communication and managerial skills are a plus in academia, and in almost any other career you might pursue. If you aim to end up in industry, however, you need to garner additional people skills, particularly in relation to leadership and teamwork. Research in industry is usually more interdisciplinary than that in academia. Team members with diverse expertise work together to create, as fast as possible, products that can be commercialized.

An academic may aim to become the world's leading expert in an interesting but narrow area of scientific investigation. In industry the emphasis is on goal-directed problem solving, and researchers preferably have knowledge in many areas of research and methodology. This enables them to play varied roles on different projects. If you choose this career path, you should also have excellent organizational skills, as you will need to navigate smoothly among many projects simultaneously. Last but not least, you must have good record keeping habits; you will not be allowed in industry to use paper napkins to write notes on your experimental procedure or comments on your data!

Network, network, network!

WORK WITH YOUR MENTOR

Work with your mentor to develop a list of categories of people with whom you should be networking and your specific goals in targeting each group. You may ask some people to share their experiences to help you define your preferred career path and attain the training necessary to pursue it. Others have hiring power. Develop lists of people you or your mentor know on a personal basis in each networking subgroup. Seek your mentor's advice regarding which professional meetings would be most profitable for networking. These meetings generally provide an excellent venue for you to impress key players and gather information about available jobs.

START AS EARLY AS POSSIBLE

The main goal of networking early in your postdoctoral term is to achieve name and face recognition. This will become one of your most valuable assets when you start your job search a few years later. The key players in your field will be more apt to help you in the future if your relationship with them did not start (by mere coincidence!) at the moment you urgently needed their help in finding a job. And if they are already familiar with your qualifications, they'll be more likely to think of you when insider's job opportunities arise.

EXPLORE VARIOUS CAREER OPTIONS

Talk to people employed in different academic and nonacademic sectors about their jobs. Here are some examples of the questions you should ask:

- What made you decide to pursue this particular career?
- What do you enjoy doing the most?
- What don't you enjoy?
- What is your overall level of job satisfaction?
- What would you do differently if you had to start all over?

Do not depend on the opinion of only one or two people in each employment category; you may encounter someone who is unrealistically happy, unreasonably disgruntled, or just having a bad day. Also, be aware that job recruiters tend to provide rosy images of the jobs they're seeking to fill.

LEARN WHAT'S REQUIRED

What will it take to get a job in a particular career environment? An easy way to find out is to have informal conversations with people who have such jobs. You need to hear detailed answers to the following questions:

- What special skills do you need to perform your job effectively?
- How did you prepare yourself for job hunting?
- What do you wish you had learned in graduate school or during postdoc training to better prepare you for your current job?
- How did you acquire supplemental job skills you did not have when you started?

You can also gain helpful information about job requirements and environment by arranging visits with human resources representatives. Such visits are commonly known as informational interviews.

SELL YOURSELF

Nothing leaves a worse impression than a letter of reference that merely summarizes an applicant's curriculum vitae, perhaps ending with a boilerplate compliment such as "Johnny is quite cheerful and always greets me in the department corridors"! Still, many graduate students and postdocs find themselves in the uncomfortable situation of having to ask for references from professors with whom they have had little personal interaction. Ensuring that those around you are fully aware of what you do will make it easier for you to name references who are willing and qualified to write letters of recommendation when you start your job search. Go to lunch with professors and postdocs in other research groups. Scholars love discussing work over meals!

When a presentation on a topic related to your research is scheduled, whether in your department or elsewhere, read some of the presenter's publications in advance and try to arrange a meeting to discuss your current research and future plans. Then attend the seminar, ask good questions, and make intelligent comments. Identify the key figures in your field and approach them at scientific meetings. They are usually curious to meet young scientists to exchange ideas (however, avoid becoming a pest by taking more of their time and attention than necessary). Stop at poster sessions to discuss research in your field with presenting authors. Make constructive suggestions regarding future experiments. Propose future collaborations. Volunteer to give research presentations, even free of charge, at neighboring institutions or in other departments at your own university.

MAINTAIN CONTACTS

Maintain contact with your PhD mentor, members of your dissertation committee, peer graduate students and postdocs, and others you know well after you go your separate ways. Keep them posted on your progress and career goals on a semiannual or annual basis. Do not allow long lapses of communication with these valuable peers. You do not want to give the impression that you only contact them when you need their help.

NETWORK WITH RECRUITERS

Professional recruiters reportedly handle three times as many job opportunities as are advertised in professional journals. Recognize that recruiters often specialize in targeting personnel in specific types of careers and at particular career ranks. Find out which ones best suit your career goals. Remember that headhunters generally earn their pay by getting a substantial fraction of the first year's salary, either from the employer or from the employee. Be sure to ask about this before committing yourself to a recruiter.

Engage your postdoctoral mentor in the process

Mentors should play an important role in preparing their graduate students and postdoc trainees for future job hunting. They should help open doors and remove roadblocks. This is more likely to happen if you maintain regular communication and let your advisor know (1) that you need his or her help and (2) that you are open to receiving career advice and will not see this as interference with your scientific or personal freedom. The following are a few of the many ways a mentor can help get you job-ready:

- Help you to develop specific future goals.
- Suggest and make available resources for professional development.
- Suggest new opportunities for networking.
- Offer honest periodic performance assessments and suggestions for improvement.

Unfortunately, not all advisors keep an open mind regarding possible career choices for their graduate students and postdoctoral trainees. Some have preconceived negative impressions of certain career directions and strongly discourage their trainees from pursuing those options. If you find that your advisor is opposing your career choice, we encourage you to seek feedback and support from others and be true to your own aspirations.

Take-home messages
- Start preparing for a future job on day one of graduate school.
- Inventory the personal and professional skills required for successful job hunting. Many of these are not commonly part of a formal graduate education curriculum.
- Develop an extensive network of people who can serve as references and informants regarding job opportunities.
- Read about various career choices available to those with your graduate degree.
- Learn firsthand about specific careers from those who have tried them out.
- In contemplating your ideal job(s), consider what balance you wish to strike between work and the rest of your life.
- Dedicate sufficient time to career development.

References and resources
Burroughs Wellcome Fund and Howard Hughes Medical Institute. 2004. *Making the Right Moves: A Practical Guide to Scientific Management for Postdocs and New Faculty*. Research Triangle Park, NC: Burroughs Wellcome Fund; Chevy Chase, MD: Howard Hughes Medical Institute.

Griffin, Jack. 1998. *How to Say It at Work: Putting Yourself Across with Power Words, Phrases, Body Language, and Communication Secrets.* Paramus, NJ: Prentice Hall.

Kreeger, Karen Y. 1999. *Guide to Nontraditional Careers in Science.* Philadelphia: Taylor & Francis.

Robbins-Roth, Cynthia, ed. 2006. *Alternative Careers in Science: Leaving the Ivory Tower,* 2nd ed. Boston: Elsevier.

Sapienza, Alice M. 2004. *Managing Scientists: Leadership Strategies in Research and Development,* 2nd ed. Hoboken, NJ: Wiley-Liss.

8

DIVERSITY OF CAREER CHOICES

Many people assume that anyone with a doctorate in the sciences or engineering is destined to become a university professor. A couple of decades ago many PhDs shared this belief, and some still do, but it is far from true. The proportion of doctorates getting academic positions, particularly tenure-track positions, has been steadily declining. University professors now hold onto their jobs for much longer due to the elimination of forced retirement and the slow growth of retirement investments. At the same time, economic problems have limited the ability of colleges and universities to create new faculty positions. While the scarcity of academic positions is a major factor contributing to the change in PhD employment patterns, one cannot ignore the fact that career opportunities outside the academy have also evolved. Such opportunities have attracted the attention of many, not simply as fallback plans, but as more attractive and lucrative possibilities. One who chooses not to pursue an academic job is no longer considered a second-class citizen or a traitor to the academy. As discussed in the previous chapter, it is important to choose a job that provides you with satisfaction, particularly from an intellectual point of view, and that is compatible with your personal and family goals and values.

Jobs in academia

Let's start by considering a conventional tenure-track faculty position in a research university—a job similar, but not necessarily identical, to that of your doctoral or postdoctoral research advisor. In such a position, your responsibilities will normally include research, teaching, and academic service. You will need to learn quickly how to juggle your time and to multitask. You will often have to work simultaneously on the next day's lecture and a research grant proposal due the following week. You will have to give your utmost attention to a research seminar by a prospective new faculty member, then switch gears and chair a session of your department's admissions committee.

How you distribute your time among these sorts of tasks will depend on your particular department and college and on the stage of your career. The variation becomes even greater if we consider institutions other than research universities. The Carnegie Foundation has classified institutions of higher education into major categories depending on their research activities and the types of degrees offered; their most recent classification was issued in 2005:

Associate's colleges. These institutions, commonly known as community colleges, mainly grant associate's degrees and certificates, with no or few bachelor's degrees. They serve as gateways for students who plan subsequently to transfer to baccalaureate colleges. Associate's colleges are further classified according to size, location (urban, suburban, rural), type of support (public or private), and type of campus (single or multicampus).

Baccalaureate colleges. These are primarily undergraduate colleges that mostly offer bachelor's degrees in either liberal arts or general fields. There are also baccalaureate/associate's colleges that offer some baccalaureate degrees but mostly associate's degrees and certificates.

Master's colleges and universities. These institutions offer a wide variety of baccalaureate and master's programs and are further subdivided according to size.

Doctoral/research universities. These schools offer a wide spectrum of baccalaureate educational program and are committed to graduate education through the doctorate. Such institutions are subclassified according to the number of doctorates awarded, the number of disciplines in which doctorates are offered, and the amount of research funding they bring in.

Specialized institutions. This classification includes stand-alone medical and other health profession schools, schools of engineering and technology, theological seminaries and Bible colleges, schools of business and management, schools or arts, music, and design, and schools of law.

Tribal colleges and institutions. These are members of the American Indian Higher Education Consortium and are generally tribally controlled and located on reservations.

In research-intensive institutions, you will be required to establish and maintain a vigorous and well-funded research program. Academia offers significant freedom; what research you can pursue is limited mainly by the availability of funding. Maintaining your lab's external funding can be the source of considerable stress; you will have to dedicate a significant proportion of your time to writing new grant proposals, revising others, and preparing progress reports for the funding agency. You will also have to spend a few hours every day with graduate students, technicians, and postdocs in your lab, discussing their findings, planning future experiments, and helping them put their findings into publishable form. Moreover, you must allocate significant time to read the literature and keep up with rapid advances in your field. These administrative, managerial, and current-awareness responsibilities will gradually take you away from the lab bench that attracted you to academia in the first place. The more successful you become at obtaining grants and expanding the size of your research team, the farther removed you will become from doing experiments. You must also beware of the danger that these mounting demands will eat up time you would rather dedicate to personal enjoyment and to your family.

In addition to doing research in a large research university, you will most likely teach or coteach a course or two each academic term. You might be responsible for directing a course, a task that involves coordinating course content, assembling a team of instructors, and administering labs and exams. Your teaching activities will include undergraduate, professional, and graduate courses, each of which requires different preparation and style of delivery. You will also be responsible for training graduate students in research, either by being their major research advisor or by serving on dissertation committees. Farther along in your career you may begin to attract bright postdoctorals to your lab. Postdocs require dedicated mentoring to prepare them for their own independent research careers. Collectively, your various research advising activities represent an important contribution to training future researchers. As your career advances, you will find yourself more involved in service and leadership, both within and outside your institution. You will likely be asked to chair departmental, collegiate, or institutional committees rather than just to serve as a member. You might be asked to head a task force to design a new curriculum or serve on a national grant review panel.

One usually joins a research-intensive higher education institution at the level of assistant professor, likely on the tenure track. At this probationary stage you do not have tenure and your appointment is renewable on an annual or sometimes two- or three-year basis depending on an evaluation of your performance. In the majority of institutions you will come up for tenure during the sixth year of your appointment. The granting of tenure, accompanied by a promotion to the rank of associate professor, relies largely on asking external reviewers to evaluate your research productivity and status in the field. The quality of your teaching and service to your department, university, and discipline are also considered. If you do not pass muster, you will be given a one-year terminal appointment, during which you will have to find a new job. Subsequent promotion to the rank of full professor involves a similar review process, with more emphasis on international reputation.

Original research receives less emphasis at baccalaureate and associate's colleges. You will have a significantly greater teaching load, which will involve teaching several courses on your own. You will also be responsible for providing academic and career advice to undergrads. Most top liberal arts colleges, however, expect their faculty to do research, generally with their undergrad students, and to publish and have grant support if they are to receive tenure. Federal agencies such as the National Institutes of Health and the National Science Foundation have funding programs specifically for researchers in institutions that emphasize undergraduate teaching, and private foundations such as the Petroleum Research Fund of the American Chemical Society, the Dreyfus Foundation, and the Research Corporation also support research in undergraduate institutions. Even at smaller liberal arts schools you will probably be expected to do some research, though that might mean partnering with a nearby research institution for facilities and collaborators. The Council on Undergraduate Research, which has as its mission "to support and promote high-quality undergraduate student-faculty collaborative research and scholarship" has a Web site (http://www.cur.org/) with useful resources.

There is even diversity of careers within academic institutions of the same classification. In a research- intensive university, for example, you might have the option of a non–tenure track faculty position. Your main responsibility then could be teaching if you prefer interacting with students to doing research (for example, supervising a set of teaching laboratories), or the other way around. You would be hired on an annual contract with the possibility of renewal and could advance up the academic ladder to the level of nontenured full professor. The entire salaries of non–tenure track research faculty often have to be provided through research grants, either yours or ones to which you are appointed by other investigators. Your job could be in jeopardy if this funding goes dry, so it is wise to establish a safety net and not to rely on a single source of funding. Your full dedication to research will afford you the time needed to explore diverse sources of research support, including federal and private grants.

You must have or acquire certain personality traits to become a successful researcher in academia. You must develop a thick skin and perseverance when faced with the rejection of a paper or a grant application. You must be able to diffuse stress and to deal with workplace politics. A novice in academia often presumes the purity of the academic mission and believes that academicians rise above common negative human traits. Academia, however, is simply a sector of human society: like anywhere else, there are sometimes nasty politics and unhealthy competition. You cannot afford to ignore this human factor in your daily interactions. Carefully craft your expressions of opposition to any prevailing dogma or to anyone else's opinion. You must learn how to deal with the malcontent and the pessimist, as much as you enjoy the company of the ever happy and the optimist.

Many academicians we have consulted while writing this book strongly advise graduate students and postdoctorals who are planning to pursue academic jobs to ascertain whether they are sufficiently committed to science to put in long working hours, both in the office and at home. They also emphasize that those who seek academic jobs must be able to multitask, have excellent spoken and written communication skills, and exhibit a healthy attitude toward an exciting career that provides not only success and glory but also intermittent setbacks.

Jobs in industry

Industry has long been an attractive alternative to academia. Progress in industry naturally depends on developing new products and technologies, which necessarily entails research. Companies direct their resources toward a wide range of goals to which you might wish to apply your research training: from discovering a new drug for the treatment of diabetes or developing an efficient apparatus to deliver a general anesthetic to building a clever robot to collect geological samples, designing a faster computer chip, or improving the aerodynamics of a rocket. Pursuing tangible goals that promise to improve human life and serve society is naturally quite gratifying.

Other advantages of a research position in industry include generous pay and benefits (relative to academic) and the assurance that you will be able to attend several scientific meetings every year. The higher salaries, however, are associated with stricter accountability than in academia. It is a common misconception that scientists in industry are free to leave at 5:00 p.m. every day and forget about their project until morning. But problems do arise late in the afternoon that require researchers to stay late. The sequence of particular experiments sometimes bring them to the lab over the weekend or a holiday. And as in academia, researchers must keep abreast of scientific progress in their field. Sometimes the only opportunity to read the literature or write an urgent report comes at home.

Research in the private commercial sector is geared toward creating marketable products. If you take such a job, you will have to bear this fact in mind in order to succeed. The top priority is to speed up the process of discovery, and along the way you will often find that you must ignore provocative research questions or intriguing findings. In this environment, time is money. There are competitors out there with similar ideas who would love to beat you to the finish line and claim the relevant patents. The emphasis on speed requires developing better approaches to problem solving and utilizing internal resources to the max. Whereas in academia you are free to contact others, outside of your lab and university, to help out when a formidable problem arises, in industry you must keep company secrets under tight wraps. This particularly applies to what you share with others at scientific conferences and what you are allowed to publish in peer-reviewed journals.

On the positive side, company management will usually provide you with the best tools to help you reach your common goals. You will have a generous budget to purchase expensive reagents and the most modern equipment. Better yet, you will often work closely with a team that has a wide spectrum of complementary expertise. For instance, a drug discovery team in a pharmaceutical firm usually consists of chemists who synthesize new compounds, pharmacologists and biochemists who screen the compounds for biological activity, toxicologists who assess potential side effects of target compounds, and pharmacists who solve problems of solubility, taste, and instability and determine drug pharmacokinetics in various body compartments. Similarly, in designing the configuration of a satellite your engineering expertise must be complemented with expertise in material science, physics, and computer simulation.

These multidisciplinary endeavors necessitate your understanding the terminology and jargon specific to each of these areas. You should also learn to appreciate the various approaches undertaken by different team members to solve a particular problem, and how they work together to streamline progress and weed out tasks of secondary importance. This powerful team structure represents an advantage that industry has over most academic settings, where research projects often have single investigators or a group of investigators with somewhat similar training backgrounds. Your success in industry will depend on your ability to be a team player and to share your original ideas and any ensuing gain with others for the sake of the company as a whole.

Throughout the process of discovery and product development you will have to maintain frequent dialog with scientific managers and marketing experts in the company. You should be able to provide convincing arguments in support of your research project and its relevance to the commercial goals of the company. In some instances, unfortunately, this might not be enough to sustain your project, or even your job. Sudden changes in market demand for the product under investigation might emerge, perhaps due to release of a similar product by a competitor. Dangerous side effects of your pet drug or fatal flaws in the wing assembly of the rocket you have designed might suddenly unfold and put a screeching halt to your project. Furthermore, you might be asked to stop working on a project you have been vested in for years to pursue a new one delegated to you by management.

More seriously, management might suddenly decide to terminate a certain line of research and close its division. They may have decided that other research projects or directions will be more profitable for the company, or the company may have been acquired by another that has different goals. In addition to the difficulties connected with loss of income, researchers may then have to relocate to find another job; both of these changes can impose significant burdens on you and your family. The extent of this vulnerability naturally depends on the economic health of the company, particularly where the company stands vis-à-vis its competitors and whether its life depends on one or a number of products in development.

There are many examples of scientists working for industry who have followed a research idea through to its realization as a marketable product a decade or two later. Others, however, have been forced to move around between projects and companies. Thus, you face some difficult choices if working for industry is your career goal. You should study the history of your potential employer carefully before signing up. Choose a company that offers decent severance packages and job placement assistance in case you lose your job. Give preference to a company located in a region that is rich in similar industries. Most importantly, stay marketable. At the end of the day, only you and your family can accurately determine your comfort zone in balancing financial rewards, risk, and what you enjoy doing when comparing industrial and academic careers.

Research jobs outside academia and industry

Research positions are not limited to academia or industry. A great deal of research takes place in state and federal agencies: state departments of health and transportation, federal agencies like the National Institutes of Health and the National Aeronautics and Space Administration, even the Department of the Army and the Department of the Navy—all of these bodies have large intramural research programs. There are also private and public research institutes that are not part of universities, for example the Cold Spring Harbor Laboratory, the Earthquake Engineering Research Institute, Argonne National Laboratories, the Scripps Research Institute, and the Vollum Institute. In each case, your research funding will be generously provided by, or supplemented with, intramural funding mechanisms that are less competitive than the grant processes you would have to rely on in academia. This will certainly allow you more time and a greater degree of freedom in your pursuit of new knowledge. You could have the luxury of asking daring high-risk, high-gain questions. The downside is that usually you will not have an opportunity to teach or train graduate students. There is sometimes a simple solution, however. A neighboring academic institution may be pleased to hire you as an adjunct faculty member or to have you lecture in their courses. You might also apply for membership in a graduate program, which would allow you to advise graduate students.

If you happen to be especially creative and adventurous, you might think of starting your own research company. You might not want to jump into this right away, though. Better to wait until you gain some research and managerial experience in an industrial setting; this will give you a chance to learn the ropes and assess the market needs. It will also give you time to arrange for initial funding and to find partners. It can be helpful to partner with someone with expertise in business development. Bear in mind that you can always start your company on a very small scale, then gradually expand. One of our favorite success stories is Research Biochemicals International (RBI). This company was established in the early 1990s to provide a handful of biochemical reagents that were then available only upon request from particular pharmaceutical firms. While the companies provided these reagents free of charge, the process was relatively tedious and involved significant paperwork and delays in receiving the material. Many scientists wished that these reagents were available for purchase from a commercial source, and RBI was the answer to their prayers. This small start-up initially distributed just a handful of such reagents, then used a very smart consumer-based approach to expand. Company representatives frequently contacted scientists to ask what other reagents they wished the company made, then produced the requested reagents. Within a few years RBI became a biochemical reagents giant; it was subsequently acquired by another giant, Sigma-Aldrich Chemicals.

Other job choices

In her book *Alternative Careers in Science: Leaving the Ivory Tower,* Cynthia Robbins-Roth lists twenty-two ways you might apply your research training away from the laboratory bench. These include being a science writer or publisher, a science broadcast journalist, a patent agent, or a regulatory affairs officer. Robbins-Roth observes that the key to success in these diverse professions is training in scientific investigation. Such training offers the analytical thinking necessary to tackle a new area. She notes apparent deficiencies in the performance of others who approach such jobs without a science background. Robbins-Roth once held a research job in a biotech company that specialized in developing novel basic research reagents, but she soon realized that science for science's sake was no longer satisfying to her. Her new goal was to contribute to the discovery of new therapeutic agents designed to combat human disease. She followed her heart and obtained a job as an advisor to the CEO of a company, working to prioritize company products according to their therapeutic potential. On the job she learned to examine intellectual property issues and to assess the competition in a thorough analytical way driven by her science training. She became knowledgeable about the process of taking a therapeutic agent all the way from the bench to approval by the Food and Drug Administration.

In the following pages we will summarize the wide array of jobs held by scientists who wrote chapters for Robbins-Roth's book. At this juncture, however, we must offer a caution. You might find one or more of the careers discussed below attractive but wonder who in their right mind would hire you for a job so far removed from your experience in the lab. You might ask "How do I find these jobs?" or "Where are they advertised?" Be patient; answers to these questions will come in the next chapter.

SCIENCE REPORTING AND PUBLISHING

Science writers attempt to translate exciting scientific discoveries into language that will appeal to a general audience. Some articles are produced for a particular magazine or newspaper, say, an article on the biology of hibernation for *Scientific American* or a report on cold fusion for the science section of the *New York Times*. Also, most research universities, medical centers, and technology institutes have their own staffs of science reporters who assemble press releases to be distributed to the media. The same is true of federal research institutions and private companies with extensive research and development programs. You could write newsletters for one of these organizations, or even compose speeches for institutional or corporate executives. Being a science writer requires dedication, perseverance, and motivation. One must also have good interpersonal skills and be accurate in reporting gathered information. Alternatively, you could obtain a job in publishing, where your tasks might include market research, product development, production, sales and marketing, or competitor monitoring.

BROADCAST SCIENCE JOURNALISM

Science journalists work in television or radio both on the air and behind the scenes. The producer of a radio science program, for example, might research background material, select and preinterview guests, and write program content. Needless to say, your research background would be invaluable in performing these tasks, increasing your ability to identify what is new and exciting in the science world, to recognize who the pioneers are in specific fields of knowledge, and to ask scientists pertinent questions. Similarly, CNN and other networks sometimes hire scientists to report on recent advances in medicine or the discovery of a new atomic particle. If you enjoy your experience as a science reporter you could advance up the ladder to become an editor who assigns stories to reporters. There are also nonjournalistic jobs available to scientists in the television world. A crime show concerned with authenticity must have, for example, forensic toxicology consultants to ensure that poisonings are accurately depicted. It would be embarrassing, for instance, to show a victim of cyanide poisoning with a red face instead of the typical bluish coloration. Your research training in material science or physics could be similarly helpful to the producers of a murder mystery that turns on the logical explanation of the sequence of gunshots.

TECHNICAL WRITING

Technical writers put together large quantities of information in a user-friendly format that can be easily understood by specific audiences. Your training in the scientific method of thinking will make the work progress more smoothly, and that will be appreciated. As a technical writer you will frequently be learning about new areas of science as you are asked to write about them. Your success will depend on your people and networking skills, as interviews with experts will represent your most valuable source of information. Technical writing offers conventional 9-to-5 working hours if you are good at multitasking and time management. The profession offers good job stability and is expanding rapidly, particularly due to increased demand for Web-based help files. Another form of technical writing is grant writing. Some universities employ or hire freelance grant writers to help individual faculty members or research teams put together well crafted proposals.

REGULATORY AFFAIRS

Scientists working in regulatory affairs act as liaisons between government agencies, the industries they regulate, and consumers, seeking on behalf of a company to provide convincing evidence that its products are safe and effective. The goods you might deal with in this type of job range from agricultural products to medical devices to air purifiers. Consequently, you could be dealing with the Food and Drug Administration, the Department of Agriculture, or the Environmental Protection Agency. You could also be involved in preliminary small-scale testing of product efficacy and safety, large-scale field or clinical trials, and issues related to advertising and conflicts of interest. This job is for you only if you are a people person. You will be constantly interacting with almost everybody in your company involved in product development, testing, and marketing. You will also be responsible for preparing documents to be submitted to government regulatory agencies. This will provide you with many chances to travel and deal with people at different levels. You might even travel internationally to have discussions with marketing and regulatory affairs experts in other countries where your company intends to distribute its products. You naturally must also have good organizational skills to succeed in this career.

PATENTS AND TECHNOLOGY TRANSFER

This job involves advising your institution or company on intellectual property issues, which often necessitates educating researchers, particularly those in academic institutions, to look for commercial opportunities related to their scientific findings. When an investigator discloses her or his discovery to you, your first task is to evaluate it and determine if it represents a patentable invention. Those who represent companies and universities in dealing with the U.S. Patent and Trademark Office are required to distill information provided by various sources within their organization into one comprehensive, and comprehensible, document that precisely describes an invention and its features. Your scientific background will be of use in collecting, and more importantly, analyzing, this mass of information. Following the patent office's evaluation of the document, you will argue for the merits and uniqueness of the invention and work to meet the requirements for obtaining a patent. Negotiation and communication skills (both oral and written) are a must in this job. You will be required to understand and speak the language of the discoverer, patent lawyers, and government administrators. This is no small feat.

Technology transfer following the granting of a patent entails making patented inventions or discoveries available for use. The job of technology transfer officers working for the inventor's institution is to identify and negotiate with industrial firms that are interesting in developing the invention into commercial products. Business developers working for industry handle the company's side of these licensing negotiations. A research background can be helpful to those on both sides of the table.

SCIENCE POLICY ADMINISTRATION

Exponential growth of science and technology and their impact on the society has generated a need for scientists who work at the interface between science and national public policy. As a result, dealing with the federal government has evolved from a task delegated to ad hoc academic consultants into a specialized career of its own. Science policy jobs are concerned not only with ensuring the responsible conduct of science but also with the proper application of scientific discoveries to social needs and the minimizing of potential harm. Practitioners' responsibilities thus range from analyzing current policies and suggesting changes to interacting with advocacy groups to resolve conflicts between science and society. Science policy administrators also prepare reports for legislative bodies on particular policy issues. To succeed in this field, you must be an eloquent writer and speaker with great analytical and people skills.

RESEARCH FUNDING ADMINISTRATION

A major stressor in academic research jobs is the relentless pursuit of external research funding. This has prompted many researchers to move to the other end of the table and work for research funding agencies. A common starting point is to become the administrator of a grant peer-review panel, for example at the National Institutes of Health, the National Science Foundation, or the Department of Energy. In such a position, you will assign research proposals to particular panel members for review, administer the review sessions, and draft summaries of reviewers' critiques for submission to the applicant. Later in your career, you might become director of one of the divisions within the funding agency. In this capacity you would deal with Congress to ensure allocation of federal funding to specific areas of research. A great advantage of this type of career, apart from the obvious one of never having to write grant proposals, is that it entails staying up to date on cutting-edge science. You will also feel immediate gratification by contributing to decisions shaping the future directions of scientific research. You should, however, be prepared to hear from disgruntled applicants whose proposals did not get the nod.

SCIENCE CONSULTANT TO INDUSTRY AND INVESTORS

CEOs in industry frequently lack scientific expertise or even experience in interacting with scientists. They rely instead on outside consulting firms to help them with specific projects. Venture capitalists who would like to invest in an exciting and hopefully profitable scientific product have similar needs for guidance. Working as a science consultant involves, in a sense, selling your scientific expertise. If you enjoy the experience you might think of eventually becoming a venture capitalist yourself. As such, you would first raise capital to create a fund, then look for opportunities to invest it using your science background to find the best investments. In doing all of that you will need to develop contacts with potential investors who are interested in scientific ventures.

MARKETING

A persuasive marketing plan requires a solid understanding of the product to be marketed. And who better understands the ins and outs of how a product works and what it does than someone with a research background and a doctorate in a closely related field? Marketing starts within the company that commercializes the product; in this context, you will normally discuss the product's special characteristics and intended uses with a team of business managers in order to produce a marketing plan. Your role could stop there or could continue to interacting with potential customers to persuade them of the product's virtues. Your scientific knowledge of customers' needs and the type of work they do will be invaluable in this regard.

We hope that we have succeeded in demonstrating that there are endless opportunities to apply your research experience to a wide spectrum of jobs. If you plan ahead and prepare yourself adequately, as discussed in the next chapter, the choice is yours.

Take-home messages
- Look at jobs both within and outside of academia.
- Investigate what it takes to succeed in careers you find intriguing.
- Find out how you can get hands-on experience that will better qualify you for a given career.
- Ask people how they got their jobs, what they like about those jobs, and what challenges they face.

References and resources

American Association for the Advancement of Science, career Web site. http://sciencecareers.sciencemag.org.

American Geophysical Union. "Careers in Science." http://www.agu .org/sci_soc/careers.html.

Find science jobs and career information. http://www.sciencejobs.org.

Fiske, Peter S. 2001. *Put Your Science to Work: The Take-Charge Career Guide for Scientists*. Washington, DC: American Geophysical Union.

Goldsmith, John A., John Komlos, and Peggy S. Gold. 2001. *The Chicago Guide to Your Academic Career: A Portable Mentor for Scholars from Graduate School through Tenure*. Chicago: University of Chicago Press.

Lovitts, Barbara E. 2001. *Leaving the Ivory Tower: The Causes and Consequences of Departure from Doctoral Study*. Lanham, MD: Rowman & Littlefield.

National Academy of Sciences, Committee on Science, Engineering and Public Policy. 2000. *Enhancing the Postdoctoral Experience for Scientists and Engineers: A Guide for Postdoctoral Scholars, Advisers, Institutions, Funding Organizations, and Disciplinary Societies*. Washington, DC: National Academy Press.

Robbins-Roth, Cynthia, ed. 1998. *Alternative Careers in Science: Leaving the Ivory Tower*. San Diego: Academic Press.

TOOLS FOR SUCCESSFUL JOB SEARCHING

Strategy and attitude

As your research training reaches the point where you are ready to move on, you are bound to hear from others how difficult it is to get a job. You will hear about the good old days when jobs were plentiful. You will learn from peers about their frustrations in finding—or not finding—jobs. This can hit close to home and might discourage you. However, there are two fundamental facts that you should bear in mind. First, there are *always* job openings. Employees still get promoted, quit, retire, or are fired. New jobs are constantly being created to meet evolving educational and market demands. Second, candidates with outstanding experience and accomplishments often get the jobs of their dreams. These basic truths about the job market should translate into two critical goals. First, you must learn how to find out about job openings, using conventional and unconventional means. Second, you must work as hard as you can to join the elite group of job candidates whom every employer wants to hire. In essence, use your fear to help you prepare rather than make you discouraged.

According to Richard Bolles, author of *What Color Is Your Parachute?*, job hunting is a learned skill that can be studied and mastered. Learn from advisors, peers, friends, and family members about their job-hunting experience. Distill what you learn to develop your own strategies. Bolles also asserts that your attitude—how you come across to people—is the single most important factor in your success in getting a job. Your attitude reflects how excited you are about a particular job and how you interact with others. It also tells about your energy, dedication, and character.

Job hunting is a process with four main steps: preparing yourself, scouting the field, deciding which quarry to go after, and embarking on the hunt. Hopefully you have read Chapter 7, which discusses how to prepare, and Chapter 8, which talks about the diversity of career choices in scientific fields. These chapters should have prepared you to decide what kind of job you would like to capture. In this chapter we will discuss how to proceed effectively on the hunt.

Think about the process from the perspective of an employer. What would you look for in a prospective employee? How would you pick finalists from among dozens of applicants, and the best candidate from among the finalists? Most people conclude that their top pick would be the person they think is best qualified to fill a gap, satisfy a need, or solve an existing problem. Every employer would also like to hire somebody who is dedicated, motivated, ethical, creative, cooperative, pleasant, flexible, well organized, goal oriented, and loyal. Now adapt these conclusions to your strategy for marketing yourself. Be honest. Do not claim traits you do not have. You should also take an inventory of the skills you can bring to a job. (But don't confuse personal traits or habits with skills.) Transferable, or functional, skills include not only the research methods you can master, but also your ability to communicate, teach, manage, and mentor.

Do not be discouraged by initial failures. Be persistent and keep trying. Never put all your hopes on one particular job opening to the detriment of others. Pursue many possibilities at once. And do not travel this bumpy road alone. Form a support group of peers and friends that will enable you to exchange experiences, ideas, and strategies.

Resources for academic job searching

Advertising in professional journals and newsletters remains the most widely used method of disseminating information about available jobs in academia. Many large academic institutions run job ads in major science journals such as *Nature* or *Science*. Others prefer to advertise in more specialized journals. You should regularly screen both types of publication. There are also some very useful online resources, including the following:

- Chronicle of Higher Education, http://chronicle.com/jobs/
- Academic Position Network, http://www.apnjobs.com
- Academic360.com, http://www.academic360.com
- Academic Employment Network, http://www.academploy.com
- American Association for the Advancement of Science, http://nextwave.sciencemag.org

Read every ad that could potentially fit your experience and career goals. If you are a biochemist, for instance, do not look only at jobs in biochemistry departments. Your specific research training could perfectly fit the requirements for a job opening in a department of pharmacology, bacteriology, or cell biology. Likewise, your experience in studying the mechanics of blood flow in a college of engineering might get you hired in a medical school department of physiology. Having a PhD should not keep you from considering employment possibilities in clinical departments in medical schools. These departments often hire PhDs to do basic research that complements ongoing clinical research and practice. The interdisciplinary nature of modern research necessitates this mix of expertise.

The ads will give instructions about whom to contact and what information to send. Applications for an academic position are usually reviewed by a search committee, so the contact person will usually be either the chair or staff person for the search committee or the chair of the hiring department.

Resources for job searching in industry

Specific job openings in industry are usually advertised in professional journals or newsletters and on companies' Web sites. Common advertising venues include *Science, Nature,* and *Chemical & Engineering News.* Interested candidates are instructed to send their résumés, accompanied by a cover letter, to a personnel office, which then forwards them to the manager of the hiring unit for screening and ranking. Top candidates are often interviewed by phone prior to selecting a few to invite for on-site interviews.

Industry also regularly uses open ads that list general hiring areas within a company. This is a generic form of advertising that lets scientists, consumers, and stockholders become aware of the company's research activities, commercial interests, and areas of growth. These open ads also serve to comply with Equal Employment Opportunity Commission requirements. Replies to these fishing expeditions are added to a résumé repository for possible future use as the need for new hires arises. Most often, however, getting a job in industry starts with word of mouth, a telephone conversation, or a face-to-face encounter. These personal exchanges may take place, for example, at a national or international conference where you present a poster. Company scientists might stop by to ask how your research technique or approach could help them solve a problem or develop a product. Their being impressed by your knowledge, willingness to share information, and clarity of presentation could be the first step toward having the company pursue you with a job offer. Company representatives also commonly participate in job placement sessions held in conjunction with conferences. This setting provides an opportunity to circulate your résumé among potential employers and to arrange for mini-interviews on location. Your participation in these events may require paying service fees, but the chance to meet people and market yourself face to face is well worth it. Some companies rely on referrals from directors of reputable academic research groups to identify potential job candidates. Be sure to let your research advisor know if you are interested in a career in industry, in case he or she is asked to recommend someone to fill a vacancy.

Industry also often relies on the services of professional recruiters, or "headhunters." These recruiters use a variety of means to locate prospective candidates, including searching their databases and public database domains, and networking through word of mouth. It is therefore a good idea to post your résumé online (e.g., at http://www.careermag.com). Companies usually have contracts with specific recruiting agencies that specialize in the relevant employment category. There are two main types of agreements between companies and headhunting firms. The first requires the company to pay a single recruiting firm set fees for conducting the search process. In this case résumés are to be submitted to the company only through the contracted recruiting firm. The second type allows the company to work through a number of recruiting firms, with payments made only to the firm that finds the candidate selected for hire, and only after that person signs a contract.

Recruiting agencies also help individuals find jobs by matching them up with potential employers. They can help you, but it is important that you find an agency that is able to understand and assess both your qualifications and the industry. Some professional recruiters specialize in serving specific sectors of industry while others have a wide scope. Here are a few examples.

- The Lucas Group serves the biotech and pharmaceutical industry but also manufacturing and engineering (http://www.lucasgroup .com/medical).
- American Biotech Associates caters specifically to the pharmaceutical and biotech industry (http://www.abapharmaceuticalindustryhead hunters.com).
- FoodHeadhunters.com specializes in recruitment for the food industry, including chemists, plant engineers, and product development managers (http://www.foodheadhunters.com).
- Russell and Partners serves the telecommunication and emerging technologies industries (http://www.russell-partners.com).

To find many other recruitment agencies and determine which best fits your needs, you can also consult HeadHuntersDirectory.com (http://www .headhuntersdirectory.com).

Additional tips for discovering job opportunities

You should not rely solely on published job advertisements, regardless of the type of job you are looking for. Word of mouth is quite effective in learning about job opportunities, particularly those that are expected to become available and advertised in the near future. For example, you might hear from your research advisor or from a colleague you meet at a research conference that a given department or industry is planning to replace a departing or retiring scientist. Try to find out details, especially the targeted research area. The potential job opening might happen to be in an area close to your expertise and in an institution or a company where you would like to be employed. You have some serious networking to do if this is the case. Your advisor might know somebody in an organization you are interested in working for. Ask your advisor to do you a favor and inquire about the date when the job opening is expected to be advertised, explore the qualifications the employer is seeking, and, most importantly, put in a good word on your behalf. You could follow up with your own contacts, either by phone or by a visit to the department chair or division head where the job opening is anticipated. Give them an opportunity to get to know you and to see what you have to offer to their research or education mission. Educate yourself about their areas of research. Discuss how you would fit in. The impact of this courting dance is immense. In addition to advertising your research expertise, it also demonstrates your enthusiasm and creativity in pursuing your goals. It is even possible that these efforts will lead the employer to formulate the job advertisement specifically to fit your expertise and background. What an edge you would then have over the rest of the pack!

Preparing and searching for jobs that require additional training

It should be apparent after reading chapter 8 that some of the careers you might pursue would require major retooling of your résumé. Robbins-Roth's book *Alternative Careers in Science* provides some excellent examples of how this can be accomplished and serves here as a main source. A scientist who became a technical writer, for example, provides the following suggestions for gaining experience before you delve into applying for positions in this field. Try writing science articles for your local newspaper on a freelance basis. Become involved in editing a newsletter in your department or college. Learn a desktop publishing or page layout program. Look for technical writing classes that will help you to write less dry prose for a wider audience. Better yet, find out if your local community college or state university offers a technical writing certificate program. Search for an internship to get real-world experience. Most technical writing job opportunities are advertised through local chapters of the Society for Technical Communication.

Getting into science writing or broadcast science journalism requires similar preparations. Take one of the science-writing courses commonly offered by universities to learn how to write news stories and science essays. This will also give you the opportunity to interact with professional writers and editors. Read basic journalism textbooks to improve your writing style. Network with local newspapers or radio and television stations. Volunteer as a freelance writer at a local newspaper to gain practical experience and build up your portfolio. Write an article or two for free if you have to. Having published your first few articles or broadcast pieces will help you get future paid assignments and may draw the attention of headhunters. Consider enrolling in a science journalism program; for recommendations consult the National Association of Science Writers Web site (http://www.nasw.org).

You might be interested in using your research experience to pursue a career in business development. One obvious approach is to enroll in a master's of business administration program; many universities offer MBA classes in the evening or online. You might also discuss your interests with the office of technology transfer at your university and seek recommendations on how to gain training. Inquire about the possibility of shadowing somebody in that office; this experience will give you a taste, albeit not a full picture, of what this career is like. If you are already employed in industry, ask the same questions of your company's business development office, and discuss your career ambitions with your boss. Many companies welcome the opportunity to train one of their competent researchers in other aspects of company operations rather than recruiting from outside. (This attitude extends to recruiting bench scientists to the division of personnel. Combined experience in research and human resources issues would make one well qualified to screen job candidates and resolve potential conflicts between scientists and management.) Training in specific areas of technology transfer is offered by the Association of University Technology Managers and the Licensing Executives Society; a law degree, especially with a focus on patent law, would also be valuable in the arena of business development. Generally speaking, headhunters are the best source of available positions in this field.

Moving to a career in regulatory affairs involves a very similar training path. You could also get postdoctoral training at the Food and Drug Administration, the Centers for Disease Control and Prevention, or the Environmental Protection Agency, then pursue entry-level jobs in regulatory affairs in one of these agencies. If you start in bench research in industry you might find your job gradually evolving in the direction of business development or regulatory affairs. The shift often starts when you contribute to a discovery that is worthy of patenting and product development. You may find yourself moving along with your discovery from one stage of development to the next. Go with the flow and learn new skills along the way.

Jobs in the arena of science policy require many of the skills you have learned in becoming a good researcher, most importantly analytical skills. However, you also need experience in writing for general audiences. You must learn how Congress works and understand politics in general. Internships, paid or unpaid, offer the best opportunity for learning the ropes. Intern in the Congress, a federal agency, or a nonprofit organization involved in shaping public policy. Many scientific societies also offer internships in science and public policy. Volunteer for a political campaign to build a network of useful contacts. Take courses in science policy. Many professional societies would love to hire a scientist to represent their interests and stance in public policy issues. Inquire about opportunities when you attend the annual meeting of your professional society. Approach interest or advocacy groups (for example, the Environmental Defense Fund or the Alzheimer's Association) that might benefit from your research background. Search for science policy employment opportunities at the National Research Council, as it is heavily involved in preparing reports on a variety of science policy issues for the Congress. Congress and congressional agencies are also a major source of employment in the field of science public policy. A large number of scientists are employed by the House Committee on Natural Resources, the House Committee on Science and Technology, and the Senate Committee on Energy and Natural Resources.

Curriculum vitae, résumé, and cover letter

Often there is confusion regarding the contents of a curriculum vitae and a résumé. Your curriculum vitae, literally a "journey of life," should contain a summary of your past and present educational and professional activities, starting from college graduation. There is no page limit. Most job openings in academia and research organizations require a vitae. A résumé, on the other hand, is a very brief document (usually a couple of pages) that succinctly states your career objectives and qualifications. This format is often required by industry. However, larger companies often ask for a full curriculum vitae. The cover letter that accompanies your curriculum vitae or résumé should highlight why you are applying for a particular job and why you believe you are a good fit for it.

CURRICULUM VITAE

While there is no set format for a curriculum vitae, the following elements should be included:

- Name, address for correspondence, contact phone numbers, e-mail address
- Education history: degrees and conferral dates, names of granting institutions, dissertation title, and name of advisor
- Postdoctoral training (if applicable): dates, name of advisor and institution, area of research
- Work experience, including teaching and research
- Current research interests and future plans
- Summary of research techniques and methods with which you have experience
- Grants and fellowships: type of funding (e.g., individual fellowship, research grant, contract), funding source, project title, your role (trainee, principal investigator, consultant, etc.), amount, and dates
- Honors and awards
- Membership in professional societies
- Professional service: participation on committees in and outside of your institution, leadership experience, consulting activities, editorial services (e.g., reviewing research manuscripts for a journal)
- Invited research presentations: dates, titles, and names of inviting institutions
- Publications: original-research papers, book chapters, and invited review articles; include full titles, authors' names as they appear in the publication, and complete citations
- Published conference abstracts: title, authors' names, dates and place of conference, and name of the conference or hosting organization

It is OK to include papers accepted for publication or under review. You can also mention important papers that are in their final stages of preparation, but no more than one or two, especially if you have only a few papers already published or in press. Listing one published paper and five in preparation strongly suggests procrastination and poor time management. For each such paper you do include, state the planned submission date and the name of the target journal.

Do not include the following information in your vitae:

- Age or date of birth
- Ethnicity, place of birth, nationality (unless the job advertisement specifies that only U.S. nationals or permanent residents are qualified)
- Social security number
- Political and religious preference
- Marital status or sexual orientation
- Hobbies

If you are applying for an academic job that emphasizes teaching you should include a detailed description of your teaching experience. We suggest you address the following points and provide necessary documentation when applicable.

- Subjects you have taught or are prepared to teach
- Courses or workshops you attended to enhance your teaching skills
- Experience with classroom diversity
- Technology teaching skills
- Sample syllabus, course assignments, and student assessments
- Student evaluation of your teaching

You may also be asked to submit a statement of your teaching philosophy, and some search committees may request a course syllabus with sample lesson plans. Submit these materials as an addendum to your vitae.

The following online resources provide guidelines, examples, and templates of curricula vitae:

- Department of Psychology, Hanover College, http://psych.hanover .edu/handbook/vita2.html
- About: Job Searching, http://jobsearch.about.com/od/cvsamples/

RÉSUMÉ

A résumé is structured to indicate briefly how your experience matches the requirements of a specific job. Thus, you should prioritize your skills and experiences to parallel the specific set of skills required. A résumé usually starts with your contact information. Use your home address if you do not want others to know you are applying for jobs and where you are applying. Describe your educational background, starting with undergraduate education. List names and locations of institutions, the types and dates of degrees obtained, and the names of your advisors (where applicable). List honors and awards, including educational and research fellowships. Include professional societies in which you are a member.

Chronologically summarize your training and employment history, including names of employers/trainers, dates, and addresses and phone numbers of contacts. Some recommend adding a very brief summary of your accomplishments in each place of employment or training. Alternatively, you can list individual skills followed by associated work or training experiences: for example, "Experience in molecular dynamics; applied computer-assisted modeling to determine the conformation of the thyroid hormone receptor" or "Drug design and synthesis; applied computational chemistry to design and synthesize polymers with potentially exceptional adhesive properties." Be informative yet succinct. Finally, list your publications and presentations at scientific conventions. Use guidelines similar to those discussed above in relation to writing a curriculum vitae. You could either include a list of references and contact information, or state that references are available upon request. For a diverse collection of résumés and templates, see http://jobsearch.about.com/od/sampleresumes/.

COVER LETTER

Employers usually require that you include a cover letter with your curriculum vitae or résumé when applying for a job. You should structure your cover letter to convey two main messages. First, why you are interested in this particular job and this particular employer. Second, what training and experience distinguishes you from other applicants. Don't reiterate everything that's stated in your vitae or résumé. Just summarize the highlights, using the position description as a guide to what's most important. A cover letter also gives you the opportunity to talk about your attitude in working with others, approach to problem solving, and future plans and aspirations.

Nowadays it is common to ask job hunters to submit their application materials online. This speeds up communication, cuts down cost, and saves a few trees. It would be wise to convert your documents to PDF, to avoid inadvertent changes in text and formatting. Formatting your application for multiple employers also makes it easy for you to apply for more job opportunities. Be careful, however. Make certain you are sending the right application material to a given employer. A quick click of the mouse on the wrong file attachment could be embarrassing. Also, follow up with a hard copy if your application includes material that requires special representation (e.g., glossy color illustrations).

Other documentation

If you are applying for an academic position you might be asked to submit a summary of your future research directions, teaching philosophy, or both. A teaching philosophy is a self-reflective statement of how you approach various aspects of teaching, both in terms of planning and execution. It is important to carefully connect your beliefs about effective teaching and the specific strategies you have used or plan to use to implement them. Address your thoughts and plans regarding all types of teaching you might be required to do at a given setting: small and large classes, labs, and so on. A University of Minnesota Web site on teacher training suggests several questions, some of which are paraphrased here, that you might ask yourself as you develop your teaching philosophy (http://www1.umn.edu/ohr/teachlearn/tutorials/philosophy/prompts.html):

- What do you think constitutes "good teaching" (i.e., teaching that promotes learning)?
- How does what you believe about good teaching enhance, resonate with or flow from the basic content, theory, and skills required for learning in your discipline?
- What does good teaching look like in practice?
- How would your students describe your teaching?
- How do you assess student learning?
- How do you assess your teaching effectiveness?
- How have you modified your teaching in response to student feedback?
- How do you put your philosophy of teaching into practice?
- What metaphor would best describe your teaching?

The job interview

A job interview is in some ways like a blind date. Its main purpose is to let your prospective employers know more about you than the contents of your vitae. You should also use the opportunity to learn more about the employer and the job. Ideally, the interview should have a significant positive impact on your final ranking as a job candidate. You have a vital proactive role in making this happen. A wise strategy is to prepare for the interview with the attitude that you might not be the top candidate going in. You must ensure that the process leaves your interviewers utterly impressed and fully aware of your unique strengths and skills. While it is always wise to recap the highlights of what is already stated in your vitae, your main goal is to add something more. Use the interview to convince the employer that you are even better than what your vitae or résumé says.

DO YOUR HOMEWORK

Much preparation is required to ensure a successful interview. You will need time to gather information and to practice, so start early.

You should collect information about both the department in which you hope to be hired and the organization of which it is a part. Learn about its research, lead scientists, and products, if any. With the advent of the Internet, this information is only a couple of mouse clicks away. There are also online resources that compare academic institutions and industrial firms. For example, the National Science Foundation provides information on academic institutions' level of research funding, number of graduate students and postdocs and their distribution by discipline, and sources of financial support (http://www.nsf.gov/statistics/). Ask your research advisor and other faculty what they know about the department and organization where you are interviewing. Past employees or trainees of the organization may provide reliable inside information, but bear in mind that some past employees may hold grudges against their former employer.

Collect similar information about other departments within the organization that could impact your job. Inquire about the possibility of meeting with key individuals in these departments during the interview. Give some thought before the interview to how you might collaborate with these individuals and how they might benefit from your expertise.

Familiarize yourself with the job responsibilities prior to the interview. Talk with people in your network who have similar positions in other organizations. Be as specific as possible in your information gathering, both in terms of the types of duties involved and the expected time distribution among them. Be aware that this distribution varies from one organization to another depending on its type, mission, size, history, and culture. You will certainly learn more during the interview, but you absolutely need to go in primed with initial knowledge.

Tabulate the anticipated job duties, then list ways in which you are qualified to perform each of them. More importantly, think of the special talents, education, and training that you believe will make you superior to others in performing each task. You must be ready to answer the central, albeit commonly unspoken, question "Why should we hire you?"

THE INITIAL PHONE INTERVIEW

Many employers start the process of candidate selection with a phone conversation. This might be a brief informal chat or a formal interview lasting up to an hour. In either case the phone interview serves to prescreen candidates for interest, professionalism, and personality. Some questions are usually crafted to find out if you are genuinely interested in this particular job. Others aim at assessing your general attitude, work ethic, and personality traits. This is your point of entry; come prepared. One obvious disadvantage of phone interviews is that you cannot observe the interviewer's body language, which in person would give you an indication of when to elaborate on a point or clarify an ambiguous answer. A video link, which is sometimes used, solves this problem. In both telephone and on-site interviews, interviewers are prohibited by EEOC regulations from asking any questions related to marital or parental status or sexual, political, or religious orientation. At the end of the phone interview you will normally be asked if you have questions about the job or the employer. Good questions reflect good preparation and candid interest rather than ignorance.

ANTICIPATING INTERVIEW QUESTIONS

Many books on interviewing skills contain lists of questions you should anticipate during a job interview. Some recommend memorizing canned answers. We do not endorse this strategy. A given question could be asked in many different ways, and in endless combinations with other questions. Pulling together an informative and coherent answer from bits and pieces of memorized information is difficult, if not impossible, especially given the stress of the situation.

On the other hand, it would be a grave mistake not to be aware of the general types of questions asked during job interviews, or not to contemplate general strategies for addressing them. This preparation will provide you with the right frame of mind to help you navigate from one question to the next with ease.

There are five general categories of job interview questions we believe you should anticipate. You cannot expect that these questions will be asked in any particular order or that the interviewers will exhaust one category before moving on to the next. The actual scenario will depend on the flow of the conversation and your answers to previous questions.

- What brought you here? What led you to pursue this particular job opportunity? Are you specifically interested in us or are you window-shopping?
- What transferable skills do you have? Why would hiring you be good for our organization? How do your skills and knowledge mesh with and complement existing expertise? How productive will you be, and how would your presence enhance productivity of the group as a whole?
- What are your personal traits and values? Would you get along with others? Would you lead, follow, or be in the way?
- Why should you be selected over others who have made it to the interview stage, particularly if you all look about equal based on your résumés/vitae?
- Are you affordable?

For the most part, these questions target information that is only partly evident from your vitae. The experiences documented there are important. The job interview is just as important. Do your best to complete the picture during the interview and leave your interviewers wowed. Sometimes you need to be clever and steer the conversation to adequately cover these issues.

Here are some general strategies for addressing the five types of interview question.

WHY DO YOU WANT TO WORK FOR OUR ORGANIZATION?

Employers naturally feel more positive about a candidate who has a clear reason for applying to their organization or organizations like theirs. On the one hand, they do not expect you to target only one place for employment. On the other hand, they will not be impressed to hear that you applied, somewhat randomly, to "a bunch of places."

Specifically summarize the information you have gathered that attracts you to this particular organization. Try to touch on as many of these attributes of the organization as are applicable:

- National/international reputation (in research, teaching, innovation, etc.) or high ranking (research funding, quality of graduate program). Cite specific rankings if you clearly remember this information.
- Reputation as an excellent work environment (without necessarily dropping names, explain how you have arrived at this conclusion).
- Core facilities (expensive equipment, well-equipped libraries, up-to-date teaching facilities). Impress them with your awareness of specifics (e.g., a linear particle accelerator, a unique telescope).

Your knowledge of the organization will have a strong positive psychological impact. In essence, it will make your interviewers feel they are talking to someone they could relate to as a colleague, and you will collect the dividends of your careful preparation.

WHY WOULD HIRING YOU BE GOOD FOR THE ORGANIZATION?

This question will almost always be asked, albeit indirectly. The best strategy in responding is to relate your specific skills and experiences to the job responsibilities and characteristics of the hiring organization. Summarize what you know about the job duties and how your educational and training background would enable you to excel in fulfilling them. State how you plan to prioritize and juggle all responsibilities and which ones you think you will enjoy the most. Also mention what you know about the skills and experiences of the current members of the hiring department and what makes your experience complementary. Show the interviewers how well your qualifications fit the future goals of the organization.

WHAT KIND OR A PERSON ARE YOU?

It is unlikely that you will be asked directly if you have certain negative personal traits, since the "right" answer would be obvious in every case. It would also be inappropriate for you to start listing your personal strengths unless you were specifically asked to do so. A clever way to indirectly deliver this important message is to include examples throughout the interview of how you handled a tough problem or dealt with a problem coworker. Let your interviewers draw their own conclusions about your attitude and personal values.

The following are personal traits and values you would like to convey—but only if you believe you have them.

- Self-motivation and enthusiasm about your work and career
- Creativity and resourcefulness
- A collaborative attitude and the ability to work well with others
- Respect for others and for diversity
- Maturity and responsibility
- Confidence but not arrogance
- Leadership
- Honesty and integrity

WHY SHOULD WE HIRE YOU RATHER THAN ONE OF THE OTHERS ON THE SHORT LIST?

Again, this question will likely be asked in an indirect manner. The question serves more than one purpose:

- To hear again why you consider yourself a good fit for the hiring organization.
- To test for a healthy balance of confidence and realism in self-assessment.
- To test for arrogance.

CAN WE AFFORD YOU?

This question translates: Could we win both the financial and the intellectual competition with other institutions or organizations interested in you? It is human nature to be highly motivated in pursuing a target only if it is deemed reachable. Most questions of this nature develop after the interviewers have decided that you are either the candidate of choice or the first runner-up. You might be asked informally in your first interview, however, if you are also considering other jobs. The most opportune time to ask you this question is during informal conversations over dinner. You are not obligated to answer this question. However, it would be advantageous and potentially profitable for you to let it be known in general terms whether you are considering other organizations and how far you are in the job-hunting process on the other fronts. You would also be wise to mention your general approach and the criteria you plan to use in making a decision. If it is the case, let the interviewers know that their organization stands a fair chance in winning you over, and why. Also indicate, as specifically as possible, how the interview has enhanced your interest in this particular opportunity. We advise you, however, to gracefully evade questions regarding salary expectations. The first interview is definitely the wrong venue for exchanging this type of information. According to Richard Bolles, author of *What Color Is Your Parachute?*, it is too early to talk money when the interview is still at the "who are you?" or "we like you" stages. Wait for the "we must have you" stage. It is too late, however, if you put off salary negotiations to the stage of "we got you."

Some interview questions might strike you as conniving and intended to catch you off guard and extract personal secrets. Other questions might seem open-ended or aimless and confusing. Do not become nervous or paranoid. Bear in mind that those who are interviewing you, if they come from the scientific or technical side of the organization, may not have been trained in professional interview strategies. An obvious exception is the case of interviewers from personnel/human resources departments. (Top industries also train their scientists in interviewing strategies.) Do your best to understand the aim of the question and do not hesitate to ask for clarification, since the vagueness of the question might not be as obvious to the interviewer as it is to you.

It is psychologically comforting to recognize that your interviewers are usually as nervous as you are during the interview. Experience, however, has taught them to hide their anxiety. One source of their worry is the possibility of losing you to another organization. They are also under significant stress as they are trying to find out as much about you as possible in order to make the right hiring decision. You are not on trial during an interview. Rather, you are supposed to be interviewing them as much as they are interviewing you.

MEETING WITH THE SEARCH COMMITTEE
You only have one chance to make a first, and hopefully positive, impression. Dress professionally. Do not use excessive makeup or cologne. Be on time. Treat everybody you deal with before or during the interview with courtesy. Never underestimate the impact of comments made off the cuff by office staff regarding the way you handled yourself. Offer a firm but not painful handshake. Maintain good eye contact but avoid gazing into anyone's eyes. Give everybody around the table their fair share of personal attention. Do not slouch or fidget in your seat. Do not speak too softly or too loudly. Do not mumble. Maintain a two-way conversation. Provide short but thoughtful

answers to questions. The rule of thumb is that your answer to each question should not exceed a couple of minutes. Otherwise, you might notice some glazed eyes and infectious yawns around the table. If you feel that you need more time to elaborate on an important point, you can always stop at the two-minute mark and ask if they would like you to elaborate further. Sustain self-confidence but avoid arrogance. Do not talk negatively about people you have worked with, even if prompted. Maintain a sense of humor, without appearing to be a clown. Get your interviewers involved in the conversation. Ask if your points of view and plans fit with theirs and the culture of their organization. This approach has many advantages. First, it will indicate that you are a team player who cares about the opinion of others. Second, it will keep the atmosphere of the interview stimulating. Third, it will tell you if you are on the right track in answering questions. Fourth, it will give your audience an opportunity to ask for more elaboration on statements that could be misinterpreted. Fifth, and most importantly, it will indicate your interest in gathering information about their organization, which will be interpreted as genuine interest in this specific job opportunity.

Politely take the lead if you sense an uncomfortable silence. Give the lead back when your interviewers become more interactive. This requires finesse; you do not want to appear rude or controlling. Learn the art of polite interruption. Use body language to indicate when you are ready to provide an answer to a lengthy, seemingly endless, question. You cannot afford to let time run out without getting your fair share of time to sell yourself. Use your peripheral vision to read people's faces and body language throughout the interview. It will readily tell you how well you are doing. Nods are encouraging; rolling eyes are not. Yawns are a terrible sign of boredom. Take immediate corrective action to get things back on track. Listen carefully to what each person says. This will not only help you address questions appropriately, but will also tell you a lot about where your interviewers are coming from, their priorities, and the culture of the institution.

Toward the end of the interview you will undoubtedly be asked what questions you have for the committee. This is an opportunity not only to gather information you need to make an informed decision if offered the job, but also to serve many other purposes:

- The nature of your questions would reflect on your maturity and depth of thoughts.
- The main theme of questions you choose to ask, and their order, would indicate your priorities. For example, asking more than once

about collegiality in the department/organization will clearly indicate that you value respect for others and team spirit.

- If you relate your questions to specific knowledge you have gathered about the hiring organization, you will leave a positive impression as someone who is there because of keen and serious interest in the position.

During this important segment of the interview, your questions must not be mostly about what the hiring organization has to offer you. Rather, you should ask questions related to the following points:

- Expectations. Questions in this area indicate that you are a responsible person who would like to know the specifics of what you are getting into. If you have not already done so, briefly comment on how you see yourself fitting into these roles.
- Roles and responsibilities of different members of the team and the hierarchy of the overall organization.
- Future goals of the organization and plans to get there. This points to your vision and potential leadership. Comment briefly on how your expertise and enthusiasm would be helpful in accomplishing the mission.
- Employee career development programs. Elaborate on your future goals and additional skills you would like to acquire to get there. This will strongly suggest that you value self-improvement and will likely not become stagnant and technologically outdated.

Again, do not inquire about salary or start-up funds. You are still in a very vulnerable position. Leave these questions for the time when the hiring authorities are certain that they want you. Only then do you have negotiating power. In fact, your interviewers, who are likely senior members of the hiring department, might be people you should avoid discussing salary with at all cost. In many instances, a salary offer for a junior but sought-after person is not too far from that of senior members of the group, due to changes in market value of new hires over the years.

INTERVIEWING WITH YOUR FUTURE PEERS

It is possible that following your meeting with the search committee you will have a chance to meet with some of your future colleagues. In most cases, these individuals will not be members of the formal search committee. Nonetheless, they are important constituents whose opinions and comments will count in the overall process. The main goal of these one-on-one meetings is to inform you in more detail of the professional interests and duties of various individuals in the organization. Of course, this is also an opportunity for more people to check you out. Here are some strategies to take full advantage of this opportunity to further market yourself.

- It is likely that questions will be repeated as your day progresses. Don't appear to lose interest; give each questioner your fullest attention. Answering the same question a second time will give you the opportunity to make your answers more comprehensive and to add things you forgot to mention earlier.
- You will gradually develop a more complete picture of the needs of the organization and what type of an employee is desired. Tailor your answers accordingly. Take every opportunity to illustrate how your unique portfolio of education and training could fulfill these needs.
- Be prepared to give a concise and well-structured answer to vague questions such as "Why don't you tell me about yourself?"
- Make time to ask individuals about their specific roles and responsibilities in relation to the general operation. Ask about what aspects of their job they enjoy and what they would change to achieve a smoother operation. Whenever an opportunity arises, suggest possible improvements—for example, a clever experimental approach to a difficult research problem or better marketing strategy—and highlight ways your specific skills and expertise would contribute to bringing them about.
- Be prepared to ask informed questions about the department or organization during your meeting with each individual. If possible, avoid asking everyone the same questions. Instead, follow up on information obtained during your meetings with others. Do not get uptight if you run out of new questions toward the end of the day. It is better to repeat questions than to state that all of your questions have been addressed. That would suggest boredom and lack of genuine interest in the job. Worse yet, it might hurt your interviewer's feelings by suggesting that your meeting was not stimulating.

INTERVIEWING WITH PERSONNEL OFFICERS

This kind of interview is most common in nonacademic organizations, particularly in industry. It is often conducted by human resources specialists who are highly experienced in human psychology and group dynamics. In some cases, particularly in industry, you may be surprised to find that one or more of the interviewers from the personnel department is also well versed in your area of scholarly studies. Clever organizations cross-train their recruiters and hiring specialists. Some personnel departments even recruit researchers and train them to become double-edged human resources specialists.

You will usually be asked a wide range of questions, moving back and forth among various topics. Some questions might focus on your work habits and ethic, while others deal with what you enjoy doing outside the work environment. Be brief. Most important, be truthful. This is a test of realistic and honest self-evaluation. Your answers to different questions will be compared to each other and contrasted with how your references describe you. It is also a test of your ability to work with a team, deal with authority, and handle stress. Again, remember never to bad-mouth others you have worked with.

This interview setting usually provides an opportunity for your interviewers to ask about your salary expectations. Again, gently defer this question; avoid even stating a desired salary range.

INTERVIEWING WITH THE DEPARTMENT CHAIR OR GROUP LEADER

Your meeting with the department chair or group leader is likely the most important of all, as this person has a critical role in making hiring decisions. It is common for a search committee to provide the department chair with a ranked list of finalists for the job. In other instances the chair asks for an unranked list of the top three to five candidates, which allows the chair more freedom in making the final choice. Thus, you must be sure to prepare adequately. Practice giving a succinct wrap-up that addresses the five important questions discussed above. Again, some of these questions will be asked indirectly. You will need to be fully engaged to read between the lines and accurately guess the type of information being sought. You should also be prepared to comment on the strengths of individuals you have met with, the quality of the facilities, and your overall assessment of the organization. Address these issues with a healthy mixture of candidness and tact: focus on strengths; state weaknesses and deficiencies tactfully. A clever way to do this is to ask the group leader what plans are in place to strengthen whichever aspect of the general operation you are concerned about. This approach will also give you hints about the leadership style of your future manager.

The interview with the group leader is definitely the most appropriate time for you to inquire about roles and responsibilities of the job. Make sure you end up with a clear picture of what the various duties are, their relative importance, and how you will be expected to divide your time among them. You should also inquire about the mechanisms and metrics that will be used to gauge your job performance. Ask about expectations for your career progress five or ten years from now and how the organization will help you get there. This type of question leaves a very positive impression that you plan to continue advancing in your career for many years to come.

At the end of your meeting with the group leader, you will usually be briefed on the anticipated timeline of the hiring process. If not, it is perfectly appropriate for you to ask. If your references have not already been contacted for letters, you may be asked if it would be OK for the organization to do so. It is wise for you to give the group leader the liberty to contact additional people in your current department. This will leave a positive impression of your self-confidence.

INTERVIEW TALK

If you are interviewing in academia, and sometimes in industry, you will be asked to give a research or teaching presentation, or both. This step gives the hiring organization a chance to assess the quality, approach, relevance, competence, and competitiveness of your research and the effectiveness of your teaching skills. Additionally, you might be asked to present the outline of a research grant proposal to demonstrate your potential to bring in external funding. This will likely be the case if you are applying for a tenure-track faculty position at a top research institution. Prepare the contents and flow of your various presentations bearing in mind their different objectives. Connectivity, smooth flow, and clarity to all are crucial elements of successful job interview presentations (refer to Chapters 19 and 20 on presentations for helpful details). Most importantly, bear in mind that a job talk is different from a regular research seminar presentation. Your main goal here is to convince as many audience members as possible that you would be a valuable addition and that your background and research techniques would provide opportunities for collaboration. If you do your homework prior to your interview and learn about the research interests of department members you could even suggest specific collaborations. An important strategy is to keep a good balance between the breadth and depth of your talk. If you can convey the importance and novelty of your research to both experts and nonexperts, you will both gain supporters and demonstrate your teaching abilities. End your talk with a statement on the future directions of your research.

INTERVIEW DINNER (AND OTHER MEALS)

Dining with your hosts is an integral part of the interview process. While mealtime provides an informal opportunity to interact, don't forget that things you say or do will contribute to the overall impression you leave. Try to steer this opportunity away from being a pure shoptalk session. Present yourself as a person who has a life outside of work. Exchange information about hobbies and personal interests. This will certainly make your hosts warm up to you as a potential member of their work community. It would also show your future peers that you are the kind of a person they would enjoy interacting with, both socially and professionally. Be sensitive to the makeup of the group in selecting topics of conversation. Avoid saying anything that could potentially hurt somebody's feelings, even in humor. Take advantage of the interview dinner to inquire about what the city has to offer in terms of entertainment, cultural events, quality of education, transportation, and so on. Such questions indicate that you are serious about the job. The interview dinner sometimes takes place the evening before the formal interview meetings. This sequence gives an early opportunity to establish rapport with your potential colleagues and can serve to make you more at ease during the formal stage of the interview. This interaction becomes especially significant if the group leader is among those attending the dinner, as it can markedly reduce your anxiety about meeting him or her the following day.

The dinner interview is an opportune time for you to discuss what the new job environment, particularly the city, has to offer to your family, if you have one. After all, your family's needs and preferences will be factors in your decision to accept a job offer. One issue of paramount importance is employment opportunities for a spouse or significant other. If you have children, inquire about such things as the availability of day care facilities, quality of family health insurance coverage, and parental leave policy.

AFTER THE INTERVIEW

Remember to send a note promptly thanking the department chair or group leader for inviting you for an interview and for the generous hospitality. Send a similar note to the chair of the search committee and ask her or him to convey your thanks to all committee members.

Immediately write down every detail of your impressions, both positive and negative. Also reflect on your interview visit and comment on what you could have done better. An opportune time to do this is on the flight back home. In any case, you must do this before you embark on another job interview. You will be sorry if you don't, as this diary will serve you well in making an educated choice if you get multiple job offers.

Do not allow time to erase any of the positive impressions you have left on your interviewers. This is particularly relevant if you were the first of a handful of candidates to be interviewed over a period of weeks or months. Keep in touch with the people at the organizations who impressed you during your interview visits. Contacting them gives you an opportunity to ask questions that occurred to you after the interview and also indicates your continued interest in the job. But avoid being a pest who calls every other day to ask whether a hiring decision has been made. You can, however, be more persistent if you have another job offer in hand to which you must respond within a limited time.

SECOND INTERVIEW

If you get invited for a second interview, you are probably the top candidate, or one of a very few remaining candidates, for the job. At this stage the hiring organization will do everything possible to win you over. You will most likely meet again with the group leader and with representatives from the employee benefits department, who will initiate negotiations and try to entice you to sign a contract.

Prior to your second interview you should do more homework to be able to reasonably and successfully negotiate salary offers. Ask your advisor and the chair of your current department about the market value of positions like the one you are considering. Calculate your financial needs. Consult up-to-date published national salary surveys. Many of these are available online:

- Salary Wizard, http://www.salary.com/salary/layoutscripts/ sall_display.asp
- America's Career InfoNet, http://www.acinet.org/acinet/default.asp
- U.S. Department of Labor, http://www.bls.gov/oco/
- Economic Research Institute, http://www.salaryexpert.com/index. cfm?FuseAction=SOCTrends.Main
- JobStar, http://www.jobstar.org
- WageWeb, http://www.wageweb.com

It is generally agreed that you should not be the first to state a desired salary figure, even when asked directly. Politely turn back the question and ask your potential employer to provide a salary figure. This will give you the upper hand. Be reasonable and realistic in your negotiations. Expect the final salary figure to lie somewhere between what was proposed and what you initially ask for. It is as important to inquire about, and negotiate, the expected source of your salary. For instance, most academic departments expect their faculty to derive a significant percentage of their salaries from research grants. You need to know the specifics. Most importantly, is your full salary guaranteed by the department in case you lose your research funding? In academia, you may eventually be expected to provide your summer salary, and perhaps some of your academic year salary, from grants. When you are starting out, you should negotiate at least a couple of years of full-year salary support, until you are able to get grant funding. In nonacademic work, full-year salary support is the norm.

Remember that the cost of living, and especially the cost of buying a home, varies markedly from region to region. You must evaluate the salary offer within the context of location. Sperling Best Places (http://www .bestplaces.net/city/) is a useful online tool that offers cost-of-living comparisons for a long list of U.S. cities. Using this and similar resources, you can compare the cost of specific types of expenses (e.g., housing, transportation, taxes, food, utilities) and the value of a given salary in cities of your choice.

An item that is indirectly related to salary is fringe benefits. Interestingly, job seekers and employers alike often neglect this important item. This is remarkable given that organizations vary so widely in how much they contribute to cost of health care, life insurance, retirement funds, and so on. This variation can be even wider than that in base salary figures. Some academic institutions, for example, provide partial or full tuition reimbursement for the children of faculty members—a wonderful benefit. Ask also about vacation, sick leave, leave to care for dependents, educational leave, maternity or paternity leave, educational cost reimbursement, and sabbaticals. In industry and other for-profit businesses inquire about profit sharing, stock options, and expense accounts for entertaining clients. You will discover that some of these benefits vary significantly from one company to another. You will also find that some benefits are negotiable while others are fixed for all employees. Knowing the base salary but not the magnitude of these indirect income supplements means not having the full picture necessary to make a fair comparison of offers from multiple organizations.

You will likely be asked to prepare a detailed list of what you need to get started on the job, particularly in academia. This includes research equipment, computers, laboratory supplies, personnel, and teaching material. Do your homework. Inquire about the average size of start-up packages at your current institution and if possible others, preferably comparable in size and reputation to the one making you the offer. Faculty members in your current department, particularly recent hires, are a valuable resource for this information.

Be as specific and as realistic as possible regarding price estimates and contemplated vendors. Actual quotes and specifications would be great. Specify the intended purpose for each piece of equipment item and the percentage of time you plan to use it. If possible, indicate what you need right away and what could wait a year or two; the department chair or group leader may find it easier to come up with the necessary resources in a phased manner. Not asking for everything at once also indicates your cooperative attitude. If possible, request that some or all of your start-up funds be kept available without a spending deadline. This will give you significant flexibility to better use the money as future needs arise.

Do not forget to ask for and negotiate these additional necessary items:

- Assigned parking (do not take this for granted)
- Secretarial and accounting support
- Technical assistance
- Moving expenses
- Assistance in selling your current home (mostly limited to jobs in industry)

Negotiations usually go back and forth for a couple of rounds before a mutually agreeable package is decided upon. Never indicate at any stage of negotiation that you are going to accept the job prior to receiving an offer in writing. This offer should specifically mention every item agreed upon during the negotiations. It should also state your job duties and how you will be evaluated.

Negotiating for a spousal or significant other hire

Many job applicants are married or have a significant other who might also be looking for a job, at least in the same city if not at the same university or industrial firm. Sometimes you will be able to apply concurrently for different jobs in the same area. This is much easier if the two of you have different backgrounds or are interested in different types of employment.

Having both partners seek faculty positions in the same institution presents a challenge but is not impossible. Most institutions would be delighted to create a job opening for your partner if he or she is equally attractive. However, prepare to be flexible and maintain open communication with your partner regarding available opportunities. For example, the second job opportunity might be in a different department or college, or might be a non–tenure track appointment. Whatever the case, you should inform the search committee of your situation when they invite you for an interview. They will do their best to help your partner get a job if they are really interested in recruiting you. Many institutions offer relocation services that cover helping with a spousal hire.

Should you accept this job offer?

You might be among those fortunate and skilled people who receive multiple attractive offers. If this is the case, you should carefully weighs the pros and cons of each position. Your comparison should focus not only on the financial aspects of the offers, but also on the following features, and as many others as you can think of. The order and weight of these criteria naturally depends on your personal preferences and values.

- Stature of the organization
- Its financial stability
- Technical and intellectual resources
- Your sense of quality of personal interactions
- Career development prospects
- Potential collaboration opportunities
- Job opportunities for your spouse/partner
- Quality of education in the region
- City characteristics (safety, transportation, cultural aspects, cuisine)

There are many online resources that will help you gather information on the characteristics of the city you are considering moving to. One example is City Profile Report (http://www.moving.com/Find_a_Place/Cityprofile/).

Many people, however, do not receive multiple offers. You may send out tens of job applications, go on a couple of interviews with a long lapse between them before, and finally get just one offer. Should you accept the job even if it is not ideal for you or for your family? The answer depends on which school of thought you subscribe to. Some strongly recommend waiting for an ideal employment opportunity. Others think that your first job is essentially a springboard and that once you prove yourself you will have more leverage to find a better job in the near future.

It is difficult for us to categorically recommend one strategy over the other. Your decision must be based on how far the job offer is from what you ideally want, particularly in terms of its potential to help you advance professionally. An environment that lacks rich technical and intellectual resources will not afford you the opportunity to advance to a point from which you can readily move up. You might therefore get stuck in a job you do not like. You should also take your family situation into account. You are naturally more mobile if you are single. Moving a family frequently across the country is psychologically as well as monetarily costly. Make a calculated guess of the probability of getting other job offers in the near future, in case you turn this one down. There may be other potential employers in the area, so you wouldn't have to move even if you changed jobs. The pharmaceutical industry in the Philadelphia–New Jersey area and the computer industry in Silicon Valley are examples.

Factors to consider in your calculations include the strength of your background compared to that of those who landed at the top of the applicant pool. Your advisor or other faculty members who have served on job search committees could be helpful in providing you with an honest estimation. If possible, use personal contacts to find out informally how you ranked among all applicants for a given job. You will likely be more successful in getting information if you ask about your general ranking (top, middle, or bottom third) than if you ask for a numerical ranking. If you or your advisor know somebody on the search committee, they might offer more insight regarding specific areas in which you were thought to lack a competitive edge. If you remained in touch with people at the hiring organization, as we strongly recommended, you should have the necessary contacts to find this information. Call the head of the search committee or the department chair or unit manager. Make it clear from the outset that you are not calling to complain but rather to gather information to help you in your future job hunting. Naturally, the closer you are to the top, the more likely you are to get future job interviews and offers.

What if you do not get the job you applied for?

Do not get discouraged. Learn as much as possible from the experience in preparation for getting a job somewhere else. Reflect on all aspects of your application, most importantly the job interview. It is obvious that the search committee members were impressed enough by your qualifications and accomplishments to have invited you for an interview. Ask yourself what might have happened during the interview that prevented you from coming out on top. Identify specific aspects that you need to work on. Speak with the head of the search committee. Make an appointment for a phone conversation—do not catch him or her off guard. Again, emphasize that you are not calling to protest not being selected for the job, but rather to ask for pointers that will help you with other job applications. Express your interest in seeking candid feedback and advice.

Take-home messages
- View job hunting as a skill that can be learned and mastered.
- Think like an employer: imagine what characteristics and qualifications you would like to see in an applicant for a given job.
- Utilize all resources that advertise jobs of interest to you, and network to find out about opportunities that might not be widely advertised.
- Acquire expertise you might lack that is important for the careers you are interested in prior to applying for such jobs.
- Be aware of the differences, in content and presentation style, between a curriculum vitae and a résumé. Invest time in crafting these documents to achieve the best presentation possible.
- Make your cover letter for a given application as specific as possible to that job opportunity.
- Prior to an interview, find out as much as possible about your potential employer.
- Contemplate why you see yourself as a good match for the job.
- Prepare for the big five interview questions: Why do you want to work here? What can you do for our organization? What kind of person are you? Why should we pick you over other qualified applicants? Can we afford you?
- Recognize that every interaction, formal and informal, during a job interview visit contributes to the impression you make on your potential employer.

References and resources
Bolles, Richard N. Updated annually. *What Color Is Your Parachute? A Practical Manual for Job-Hunters and Career-Changers.* Berkeley, CA: Ten Speed Press.

Burroughs Wellcome Fund and Howard Hughes Medical Institute. 2004. *Making the Right Moves: A Practical Guide to Scientific Management for Postdocs and New Faculty*. Research Triangle Park, NC: Burroughs Wellcome Fund; Chevy Chase, MD: Howard Hughes Medical Institute.

Chandlers, C. Ray, Lorne M. Wolfe, and Daniel E. L. Promislow. 2007. *The Chicago Guide to Landing a Job in Academic Biology*. Chicago: University of Chicago Press.

Robbins-Roth, Cynthia, ed. 1998. *Alternative Careers in Science: Leaving the Ivory Tower*. San Diego: Academic Press.

CONDUCTING AND PRESENTING RESEARCH

10 THE MEANING AND RESPONSIBLE CONDUCT OF RESEARCH

In this chapter we will discuss the meaning and broad social and ethical implications of research, issues that all beginning researchers should be aware of.

What is research?

It is interesting to consider some definitions and synonyms of "research," which bring out the various implications of the term. *Merriam-Webster's Collegiate Dictionary* defines research as "investigation or experimentation aimed at the discovery and interpretation of facts, revision of accepted theories or laws in the light of new facts, or practical application of such new or revised theories of laws." That's a rather dry definition. Less formal, and perhaps getting closer to the psychology of researchers, is author Zora Neale Hurston's statement "Research is formalized curiosity. It is poking and prying with a purpose." And philosopher Jacob Bronowski writes, "That is the essence of science: ask an impertinent question, and you are on the way to a pertinent answer." These two quotes capture the somewhat irreverent attitude that characterizes the best research. Again to quote Bronowski: "It is important that students bring a certain ragamuffin, barefoot irreverence to their studies; they are not here to worship what is known, but to question it."

"Research," as a verb, has a lot of synonyms, each with a slightly different connotation: *analyze* (to break down into parts for study), *examine* (to scrutinize carefully or critically), *explore* (to probe in pursuit of discoveries), *interrogate* (to question relentlessly), *test* (to engage in experiments). Indeed, different kinds of research, and the same project at different times, may entail hunting, surveying, dissecting, critiquing, and much more.

Here's one more word that characterizes both the outcome and the genesis of successful research, which is why those who have embarked on a research career remain so committed to it: inspiration. The word means, at its root, "breathing in," and refers both to a sudden exciting thought, seemingly snatched from the air, and the exhilaration that often accompanies it.

Basic and applied research

Research is one of the most characteristic activities of modern civilization. On the one hand, it contributes to our broad and deep understanding of our world, to the very fabric of what we call "civilization." On the other, it underlies the continuing advances in technology and medicine that we take for granted. Perhaps most beginning graduate students, and most academic researchers, are motivated by the quest for understanding. They do what is commonly called "basic research." But most members of the public who support academic research do so because of its practical benefits. These applications, many of which are pursued in industrial settings—think pharmaceutical, chemical, and chip technology companies—are developed through what is generally called "applied research."

The boundary between basic and applied research is notoriously unclear. Not only is the distinction fuzzy, it is usually unnecessary for scientific purposes, though it may be useful in appealing for funding or for public support of a given line of work. (The public usually understands the need for applications, which drive economies and produce jobs, better than it understands seemingly pointless, "impractical," basic research.)

Basic research is primarily driven by the desire to explore or understand. Its motivation often starts out as curiosity about some part of the world. Because it has no immediate intended application, it is often called "pure," and because it may be the basis for more practical applications, it is sometimes called "fundamental." Theoretical research is likely to be viewed as basic. (We should acknowledge that pragmatic motivations, such as obtaining grant support to sustain a research lab, or publishing papers to achieve scientific recognition, promotion, tenure, and salary increases, generally are present even in "pure" research.) Basic research takes place over a longer time span than applied, since it is not driven by the need to produce a profitable product in a limited amount of time. Research programs in universities may go on for decades.

What is "fundamental" in one area of science may be viewed as an application of another. For example, P. A. M. Dirac, one of the founders of quantum mechanics, opined that chemistry was just an application of quantum mechanics. Needless to say, most chemists would beg to differ. In a somewhat different example, a developmental biologist in the College of Biological Sciences working on genetic control of plant growth may view herself as doing basic research, while a colleague in a neighboring department in the

College of Agriculture may be doing very similar work with the applied aim of producing a faster growing, and thus more commercially viable, crop.

Tool development would generally be viewed as applied research: how can we measure something more quickly, sensitively, accurately? But the ability to measure better, or to measure new things, underlies nearly all advances in basic experimental science, and thus is essential in constructing new theoretical ideas. By the same token, tool development depends on scientific understanding of the principles underlying the tools. Once again, we see that basic and applied science are intimately, perhaps inextricably, intertwined.

Applied research is driven by the desire to solve specific problems that may have some practical impact. It is more likely to be funded by industry than by federal grants, though in both cases it may be carried out in an academic setting. Basic research provides the foundation on which most applied research and technology is based. Investments in basic research in the United States, especially after World War II, have led to the great economic expansion that we've enjoyed. Basic research also trains most of the students and postdocs who go on to fill positions not just in universities but in industry, government, and research foundations.

Engineering science tries to bridge the basic and applied gap. To quote a statement on the Department of Engineering Science Web site at the University of Toronto, "We cultivate the curiosity, depth of critical thinking and powers of observation that scientists exploit to make new discoveries, as well as the creativity, technical capability, and problem solving skills that engineers use to bring discoveries to pragmatic reality for the benefit of society." This attitude is not confined to engineering science departments or disciplines; it is becoming more prevalent in all branches of science.

Intellectual property and technology transfer

Basic research, driven by curiosity, can have profound practical consequences. Mary Lasker, a philanthropist and advocate for medical research, said, "If you think research is expensive, try disease." Basic research on genetics, biochemistry, cell biology, and developmental biology using model organisms leads to understanding of the molecular and cellular changes that underlie diseases. Such research may lead to savings in terms both of human misery and of money spent on health care. According to the National Institutes of Health, investments in basic research are repaid ten- to eighty-fold in reduced health care costs and increased productivity due to longer life and decreased illness. Model organism studies using fruit flies, mice, and worms are leading to understanding of human growth and development, with implications for possible cancer cures.

Basic research cannot only lower health care costs, it can lead to other major economic consequences. In physics, the invention and subsequent utilization of the transistor is perhaps the most obvious example of profound social change and economic shifts coming from basic research. From the nineteenth century comes the response of Michael Faraday to British prime minister William Gladstone, who had asked of what use Faraday's experiments on electricity might be: "Why, sir, there is every probability that you will soon be able to tax it!" Returning to biology, basic studies on infection of bacteria by viruses led to the discovery of restriction enzymes, fundamental reagents of the biotech industry.

With such large practical and economic consequences at stake, concern over the ownership of patentable intellectual property stemming from basic and applied academic research has grown dramatically— among both universities, which need to increase their revenues, and researchers, who would like to profit from their discoveries. The Bayh-Dole Act, passed by Congress in 1980, was intended to encourage the commercial utilization of inventions produced with federal funding. Universities (and other nonprofit organizations) retain ownership of such inventions and are expected to file for patents and to pursue commercial licensing of the intellectual property. Such licenses may be granted exclusively to a single company, so long as it actively pursues commercialization. Exclusive licensing is often viewed as essential to commercial success and thus to encouraging companies' interest in inventions. The overall benefits to society—both financial (economic development, taxes, jobs) and through the new capabilities enabled by the invention—are thought to outweigh the fact that the invention, enabled in the first place by public funding, is not freely available to all taxpayers. In fact, such openness led to a lack of interest on the part of companies. And before Bayh-Dole, when the patent rights resided with the government, the government did not have the resources to develop or market the inventions.

As a student or postdoctoral, you should be aware of your potential rights as an inventor. If you are one of those who make a discovery or develop a patentable process, you have rights to be listed along with your faculty research advisor as an inventor on the patent disclosure. The legal requirements for being an inventor are beyond the scope of this book but can be explained by your university's patent office if you are involved in a potentially patentable discovery or invention. The United States Patent and Trademark Office has a Web site, http://www.uspto.gov/go/pac/doc/general/, that offers general information concerning patents. The proceeds from licensing of an invention typically are divided, in roughly equal parts, among the inventor(s), the inventor's department, and the university.

Patenting and licensing of inventions is just one part of the broader topic of "technology transfer," which, as its name implies, is a process whereby research findings and resources (generally federally funded) held by universities or federal labs are shared with the private sector. The hope is that the recipients will develop new industries, or new products and processes within existing industries, which will in turn enhance national economic competitiveness, create jobs and new companies, and produce tax revenues. It is worth noting that the research resources thus transferred are not just new inventions or discoveries but also access to facilities, including specialized equipment, software, and databases. But arguably the most important transfer is the education of bright young scientists, engineers, and managers, knowledgeable about the latest developments in their fields, who will staff existing companies and found new ones.

The emphasis on the financial fruits of scientific discovery can have the effect of inhibiting free communication among scientists. In some ways this is not so different from the traditional competitive stresses that lead some researchers to hide their most recent results from their competitors or to deny access to data. Things can become most complex, and most troubling, in material transfer agreements where the free exchange of research materials (especially in the biological sciences) is inhibited by restrictions on subsequent use, or claims on a share of the profits of such use if patented and commercialized. An interesting discussion of patents in biotechnology, including the benefits (and detriments) of exclusive licensing, can be found at http://www.nap.edu/readingroom/books/property/.

How is research funded?

Since research is such an important part of modern society, it receives a lot of funding, though never enough to satisfy academic researchers. Most funding for scientific research within universities comes from federal agencies, foremost among them the National Institutes of Health (NIH), the National Science Foundation (NSF), the Department of Agriculture (USDA), the Department of Defense (DOD), and the Department of Energy (DOE). According to the NSF, federal obligations for research in science and engineering at colleges and universities in 2004 totaled more than $22.3 billion (see http://www.nsf.gov/statistics/nsf05307/). Some support, mainly for applied or targeted research, comes from corporations. Additional funding, particularly for research devoted to the prevention or cure of diseases, comes from foundations such as the American Cancer Society.

Research funding is extremely competitive: NIH and NSF typically fund only 20 to 30 percent of applications, with rates in the teens not uncommon during tight budgetary times. This means that, on average, a principal investigator (PI), the faculty member in charge of a research group, may need to submit four or five grants in order to get one funded. Because grant durations are typically three years, with more than five years being very unusual, science and engineering faculty spend a large part of their time writing and revising grant proposals.

Grant applications to most federal agencies are reviewed by panels of fellow scientists. Obtaining a major grant is taken as an indication that a researcher's work is highly regarded by his or her peers. Having received stable funding is therefore often a condition for granting tenure and promotion from assistant to associate professor. In many medical schools, for example, an assistant professor must have received at least two NIH grants to achieve tenure: the first as an indication that relevant members of the scientific community think that the research program is promising, the second as an indication that the program is indeed bearing fruit. To add to the pressure, part of the PI's salary, after a several-year start-up period, is commonly expected to come from grants.

It can be argued that receiving a peer-reviewed grant constitutes more persuasive evidence of high standing in the scientific community than publishing a peer-reviewed paper. Grants are genuinely hard to get, while most papers can be published somewhere with enough persistence. Acceptance and funding rates show that only publication in the most selective journals, such as *Science* and *Nature*, is more competitive than getting an NIH or NSF grant.

Given the importance of grant support to a PI's research program, it follows that you—as a key member of the PI's research group—will be expected to play a role in producing research that leads to publications that in turn justify continued grant support, as well as preliminary results that may lead to new grants. You may also be expected to write parts of grant proposals—on the methods with which you are most familiar, some of the background literature, and the results you have obtained—as part of a team effort. This is an excellent learning experience. You may also, if you are a postdoc, be expected to write your own grant proposal for fellowship support. We discuss grant writing in chapter 25.

If you take a postdoc in a government lab or with a private company, such as a pharmaceutical corporation, the funding for your salary and research expenses will likely come from within the organization. Not having to worry so much about research funding from grants is one of the major benefits of working in such an organization.

Why should the results of research be believed?

Science is such a potent force in modern society, with such potential to upset both economic arrangements and fundamental beliefs, that the validity of its conclusions are frequently challenged. It is therefore important to devote some consideration to the nature of scientific proof: why scientific results should be believed, both by other scientists and by people at large.

The most general reason for confidence in the results of scientific research is its strong methodological base. The hallmarks of scientific research, it has been asserted, include purposiveness (being goal-directed), rigor (intellectual honesty and consistency with a set of principles), objectivity (independence from personal feelings or beliefs), and precision. Hypotheses must be testable (able to be verified or proven false), tests must be replicable, and findings must be generalizable to different but related situations. Results are evaluated in terms of degrees of confidence (probability that observed results are not due to chance), and conclusions grounded in the principle of parsimony—a preference for the least complicated explanation (http://en.wikipedia.org/wiki/Empirical_research).

In philosophical discussions, the hypothetico-deductive method (frequently called "the scientific method") is often put forward as the ideal. It consists of four steps:

- Characterization of a phenomenon by observation and measurement
- Formulation of a hypothesis or theoretical explanation of the observations
- Prediction of consequences of the hypothesis by logical or mathematical methods
- Experimental test of the predictions

These steps are iterated until, ideally, a hypothesis is developed that satisfies all the observations, makes useful predictions, and is consistent with a broad range of scientific knowledge. The hypothetico-deductive method is most applicable to the quantitative, experimental, physical sciences in which precise measurements can be made and quantitative, predictive theories can be constructed. In broad outline, it is also applicable in the more empirical sciences such as biology or geology. Although most practicing scientists realize that a rigid adherence to "the scientific method" is not how most research is carried out, these iterated steps can generally be discerned in well-regarded results.

To gain acceptance in the general body of scientific knowledge, hypotheses cannot be formulated and tested solely within a single research group (or a single researcher's head). Science is a social activity, and peer review of experimental and theoretical results and ideas is essential if the results are to become part of the broader scientific fabric. Such peer review occurs at various stages during the evolution of a hypothesis: within a research group, at lectures and conferences, in journal publications, and in grant proposals. Peer-reviewed publication is generally considered the acid test. In later chapters we discuss how each of these levels can be most fruitfully handled.

Some areas of research touch on deeply held beliefs and fears. The most obvious example is evolution. It is important to recognize that each individual piece of evidence for most evolutionary processes is fragmentary and may not be convincing in itself—particularly to a determined skeptic. So it is crucial to place the arguments and the evidence in a context of consistency with well-established observations and theories from other realms of science. In the case of evolution, those would include geology, radioactive dating, and genomic and morphological similarities among organisms. In other areas, where experimental or observational evidence is relatively skimpy, such as cosmology, reliance on well-established physical theories is required.

Some other areas of research deal with communities that may have reason to be suspicious of the motives of university-based researchers. Many such examples have arisen recently in studies of the physical, mental, or social health of poor communities. Special approaches need to be employed to gain the trust and collaboration of people in these communities, such as spending extra time to make personal connections with research subjects and community leaders, and allowing them to participate in designing and carrying out the study and execution and interpreting the results.

For centuries, philosophers have debated the basis on which scientific statements should be believed. One basis is induction, the idea that if something has occurred in all observed cases (the sun has risen in the east for as long as anyone can remember), then that something will occur in every case (the sun will always rise in the east). Of course, there is no logical reason that, just because something has always happened, it will continue to happen; so induction has its limitations.

A second idea is falsifiability, as espoused by the philosopher Karl Popper. According to this concept, if a statement or theory is to be scientifically useful, it must be falsifiable, or able to be proven wrong. If not one falsifying instance can be devised, then the theory is—at least provisionally—considered to be worthy of belief. However, ad hoc additions to a theory can always be made to save it from falsification, so this idea also has its shortcomings.

Although most scientists probably worry rather little about the formal reasons for belief in the truth of scientific statements, if pressed they would probably combine induction and falsifiability with a third concept: coherence, or consistency with the full spectrum of supposed scientific knowledge. A chemist determining the structure of a new compound by nuclear magnetic resonance would assume the correctness of the theory and instrumentation of NMR, which are based in turn on electromagnetic theory and quantum mechanics, theories that have been used effectively in innumerable other applications. As noted a few paragraphs above, the theory of evolution and theories of cosmology also depend on networks of other, well-established scientific methods and results.

Scientists generally recognize that scientific "truths" are contingent, always susceptible to being modified or overturned by new evidence or better ideas. However, many people who are not scientists have just the opposite view, that science lays claim to infallibility. This leads to cynicism and skepticism about science when previously held scientific ideas are refuted. This often happens in health and nutrition research, as, for example, when an earlier conclusion that margarine is healthier than butter was overturned because of a new understanding of the health effects of trans fat. Also, even solidly grounded scientific results may collide with other strongly held beliefs, such as those based on religion. When science and technology are claimed to be the prime arbiters of truth and value in society, the claim is likely to provoke a backlash among those who have other, conflicting, strongly held ethical or religious values. And it is important to recognize that scientists themselves hold values that may influence their research, in their choice of problems and in their intuitions about the most convincing standards of proof.

The responsible conduct of research

Science is a social activity. It is not conducted by isolated individuals working by themselves in a search for truth. Even the most withdrawn scientist depends on theories and data, on tools and technologies, developed by others. The problems to be solved, their significance and contextual meaning, and what constitute satisfactory solutions, are defined within a broad and complex social fabric of science. If this fabric is to hold together, scientists must be able to trust each other to behave honestly and responsibly.

The new knowledge obtained through the research of an individual scientist does not fully become part of the scientific fabric until it is evaluated and accepted by others. As stated in *On Being a Scientist: Responsible Conduct in Research*:

> This process occurs in many different ways. Researchers talk to their colleagues and supervisors in laboratories, in hallways, and over the telephone. They trade data and speculations over computer networks. They give presentations at seminars and conferences. They write up their results and send them to scientific journals, which in turn send the papers to be scrutinized by reviewers. After a paper is published or a finding is presented, it is judged by other scientists in the context of what they already know from other sources. Throughout this continuum of discussion and deliberation the ideas of individuals are collectively judged, sorted, and selectively incorporated into the consensual but ever evolving scientific worldview. In the process, individual knowledge is gradually converted into generally accepted knowledge.

This process of review, revision, and refinement by the scientific community reduces the influence of individual subjectivity and encourages investigators to be self-critical before their results are criticized by others. To be effective, however, it rests upon a presumption of absolute honesty and openness in the reporting of research results. Only if other scientists can rely on the veracity of the results reported, and can know enough about how they were obtained to check them by replication and to extend them, can the results be adequately critiqued and reliably incorporated into the growing body of scientific knowledge.

Scientific advances are often rooted in new methods applied to old problems, or existing methods pushed to their limits. In such cases, disputes may arise about whether the methods are adequate or appropriate to solve the problem. Are they really measuring something relevant? Has the signal been convincingly separated from the noise? Are all the potential sources of error understood and accounted for? Have proper statistical tests been applied? Under such circumstances, it is especially important for researchers to be explicit and forthcoming about the methods they used to get their results.

Conflicts of interest can lead to biases in the execution or evaluation of research, or at least to a suspicion on the part of the public that research has been biased. Certain types of conflicts of interest have always existed: desire for recognition of priority, for first publication and recognition of an important discovery, for maintaining grant support. These can lead to temptations to cut corners: to use information contained in a privileged communication (a manuscript or grant you've been asked to review), to withhold troublesome date when submitting an article for publication, or to claim to have achieved a result when in fact there are only preliminary, inconclusive data to support the claim (something you might be tempted to do if you're rushing to finish your dissertation). These malfeasances have existed as long as people have competed for successful careers in science, and the concern exists that they are increasing as the competition for positions and research support becomes fiercer, but they are handled by standard principles and practices of scientific ethics.

Other conflicts of interest, based on business or financial considerations, have come to the fore more recently, as scientific discoveries have increasingly come to be seen as sources of economic benefit. For example, a scientist retained as a consultant by a pharmaceutical company may be suspected of having a bias toward positive evaluation of a product manufactured by that company in a drug trial. A faculty member who is also the founder of a start-up biotech company may be suspected of slanting his university research to enhance the prospects of his company. Or a professor with a research grant from a company may put pressure on a graduate student to work on the company's project rather than another project that the student deems to be of greater interest.

Most universities have conflict of interest policies that deal with these situations. You should know the policies of your institution. For our purposes, it suffices to be aware that there are many sources of temptation in carrying out and reporting scientific research. It is crucial, for both personal well-being and the integrity of the scientific enterprise, to resist those temptations.

Research misconduct

When temptations to fudge research results are not resisted, we find ourselves in the realm of research misconduct—not a good place to be! Scientific fraud, like other types of legal and ethical misconduct, may sometimes seem like a tempting shortcut, but it is always wrong. Not only will it destroy the career of the perpetrator if detected, but it has broader consequences as well. Fraud breaks the link between human understanding and the empirical world, it erodes the foundations of trust, it can harm those who depend on scientific results, and it undermines the confidence of society in science.

FEDERAL POLICY

It is useful to begin with a summary of federal policy, since the great majority of scientific research in academia is supported by federal grants and contracts, and federal standards will therefore be applied to investigations of research misconduct. The policies of the Public Health Service (which includes the NIH) are typical.

Sec. 93.103. **Research misconduct.** Research misconduct means fabrication, falsification, or plagiarism in proposing, performing, or reviewing research, or in reporting research results.

(a) Fabrication is making up data or results and recording or reporting them.

(b) Falsification is manipulating research materials, equipment, or processes, or changing or omitting data or results such that the research is not accurately represented in the research record.

(c) Plagiarism is the appropriation of another person's ideas, processes, results, or words without giving appropriate credit.

(d) Research misconduct does not include honest error or differences of opinion.

Sec. 93.104. **Requirements for findings of research misconduct.** A finding of research misconduct made under this part requires that—

(a) There be a significant departure from accepted practices of the relevant research community; and

(b) The misconduct be committed intentionally, knowingly, or recklessly; and

(c) The allegation be proven by a preponderance of the evidence.

Sec. 93.222. Research. Research means a systematic experiment, study, evaluation, demonstration or survey designed to develop or contribute to general knowledge (basic research) or specific knowledge (applied research) relating broadly to public health by establishing, discovering, developing, elucidating or confirming information about, or the underlying mechanism relating to, biological causes, functions or effects, diseases, treatments, or related matters to be studied.

Sec. 93.224. Research record. Research record means the record of data or results that embody the facts resulting from scientific inquiry, including but not limited to, research proposals, laboratory records, both physical and electronic, progress reports, abstracts, theses, oral presentations, internal reports, journal articles, and any documents and materials provided to HHS or an institutional official by a respondent in the course of the research misconduct proceeding.

The National Science Foundation defines research as follows:

Research, for purposes of paragraph (a) of this section, includes proposals submitted to NSF in all fields of science, engineering, mathematics, and education and results from such proposals.

Other definitions in the NSF regulations are essentially the same as for the PHS.

Note that these definitions of research are specific to the types of research sponsored by each agency: the PHS is concerned with health-related research, while the NSF has a broader purview. Note also the definition of "research record." It includes not just published reports but also notebooks, grant applications, progress reports, and conference presentations. In other words, be honest in all your private and public recording and reporting of research results. This is an issue of sufficient importance that it should be a matter of regular discussion among faculty and department chairs/heads and their students and postdocs.

These policies identify three types of research misconduct: fabrication, falsification, and plagiarism. We'll discuss plagiarism later, in the chapters on writing. For now, let's consider fabrication and falsification. All would agree that blatant examples of such actions are reprehensible. But there are variants that, while equally wrong, can tempt even an honest scientist in the heat of the moment.

FALSIFICATION (SELECTING OR CHANGING DATA)

Falsification of data can take two forms: omission and commission. In omission, data that don't fit the expected or desired pattern are left out. There can be valid reasons for this: an instrument was miscalibrated or misread, a reagent went bad, there was a slip in a data-reduction calculation. These reasons for omitting data must be documented if at all possible, not merely assumed. This generally means that experiments should be replicated and calculations double-checked. Replication can be a problem if the experimental material was very scarce or expensive, or if an observed phenomenon was fleeting. But that is not an excuse for leaving out data that you think were erroneous, without demonstrating (on a calibration standard if the original material is no longer available) that the equipment was out of whack. Of course, this speaks to the desirability of making sure the equipment is properly calibrated before you do the crucial experiment. Likewise, if at all possible you should have some independent test of the purity or activity of your compound before you do an elaborate experiment on it.

Even if all the experimental conditions are OK, it is still likely that the data will be noisy, showing some scatter around the average behavior. Much debate rages about whether it is justifiable to exclude data from an otherwise relatively smooth plot if there are a few outliers that clearly lie off the curve. It is sometimes said that this is permissible if the points are more than two standard deviations off the curve. But be careful! If the data are normally distributed, a deviation of two standard deviations or more will occur 5 percent of the time, which is not that uncommon. If all of the rest of the data adhere tightly to the curve, then further investigation may be warranted into why those points are off. Maybe the experiment went bad that day, in which case it should probably be repeated. Maybe those samples were different, in which case the deviation may be telling you something interesting. Maybe there was a resonance under those particular conditions. There are a number of possibilities that might lead to greater scientific insight if these inconvenient data were checked out, rather than simply eliminated from the plot.

Two further points: First, identification of outliers will be much easier and more convenient to correct or explore if data are graphed as they are accumulated, rather than at the end of a long series of experiments when the sample is used up, and the equipment is dismantled, and you're tired, bored, and ready to move on. Second, experiments should be repeated whenever possible (in triplicate would be nice) and the results reported and graphed as mean plus or minus standard deviation. That gives both you and your eventual readers a better idea of how reliable the data are, how likely a deviation from the line is to be a real effect rather than an unavoidable fluctuation in the data. If replication is not possible, then careful analysis and reporting of possible sources of error is even more crucial. Consider the elementary particle physicists, who devote years of time and many millions of dollars to trying to detect a particle that may have a life span measured in femtoseconds and reveal its existence, by leaving ambiguously interpretable tracks in a particle detector, perhaps three times in a month. Statistical analysis of these experiments is rigorous in the extreme and hotly debated, since the outcome may determine the confirmation or disproof of basic theories of the physical universe.

If you—after discussion with your research advisor—have decided that there are good reasons to leave out some data, it is important to carefully explain why in your notebook, and perhaps in the manuscript or thesis that derives from it. If there is ambiguity, perhaps the best way to handle it is to include the data in the report but do two statistical fits: one that includes the outliers and one that omits them. You can then report both and state that you are using the latter for further analysis because you believe (but cannot definitively prove) that the omitted points are invalid for a given reason.

You may be tempted to change data if, for example, the measurements don't agree with your hypothesis but would if the baseline were adjusted by a modest amount. You can think of a plausible reason why the baseline might be in error, so rather than recalibrate the instrument you just make the "correction." (Of course, you wouldn't have thought that the baseline needed correction if the results agreed with your preconception.) This is an error of commission, and approaches closely to fabrication.

FABRICATION (MAKING UP DATA)

Suppose you forget to do one experiment, or the instrument breaks down with just one point left to be measured. You "know" what the result will be, so you don't bother actually to do the experiment and just put the expected point on the graph. It's tempting, but don't do it. Either repair the instrument and do the experiment, or omit the point.

Errors of commission, or making up data, are generally considered more serious ethical lapses than omitting data. Why should this be so? The obvious reason is that one is making things up out of whole cloth, constructing a total fiction. This is clearly not the way in which to maintain people's trust in science! If a significant number of researchers invented facts to support coherent but totally made-up stories, science would be destroyed. Even if only one person did, and that work was used by someone else, a damaging trail could be started, and time, resources, and peoples' careers could be damaged. In contrast, if some data points are omitted, at least the remaining points reflect reality. The distinction between serious and not-so-serious misconduct might be whether the omitted points were just noise—cleaned up so as to make the data look better and more convincing, but not really changing the trend—or whether after the points were removed a different story appeared, one perhaps in greater agreement with the presuppositions of the investigator. The distinction is a difficult one, since either change may give the resulting paper more credence than one in which the author was perhaps more honest and came to a different conclusion. If you're drawing a picture by connecting the dots, you get quite a different picture if you don't use all the dots. Shaping a scientific story by making up points and doing so by leaving out inconvenient points shade into each other in an uncomfortably close way. Don't do either.

As this section has tried to make clear, much falsification or fabrication of data is not consciously evil or devious; it can arise simply because you're psychologically so committed to your hypothesis that "adjusting" the data such that they fall into line seems entirely natural. The remedy is easy to state, though not always easy to carry out: Don't fall in love with your hypothesis. Be stubborn, but not too stubborn, in defending your point of view. Don't deceive yourself, and keep looking for alternative explanations. The best way to force yourself to consider alternatives is probably to test your ideas on others—in casual conversations, in research group conferences, and at professional meetings. It's amazing how someone else can casually ask a key question about a control experiment or suggest an explanation that has escaped your notice despite months of mulling over a problem.

We've been writing here about data handling in a single lab, by a single investigator. Increasingly, projects are collaborative; and clinical trials in particular often involve many research centers and many investigators. There are special protocols for uniform coding, transmission, archiving, and analysis of data from such collaborative studies. At least two important questions arise: What if one of the collaborators submits fraudulent data? And what if one of the members of the collaboration wants to interpret the data differently, or use it in different ways, than the others? We leave these important questions as exercises for the reader.

ADVISOR RESPONSIBILITY

It can't be overemphasized that keeping a good, complete, comprehensible notebook (or the equivalent on the computer) is the key to data integrity. We'll discuss in the next chapter some of the details of doing so. A beginning student cannot be expected to know all of these details without training and supervision. This implies a need for a degree of oversight by the lab director that not all live up to. The lab director should inspect the notebooks regularly and in detail, looking as much for what might have been left out (Joe Smith was in the lab last Tuesday and I saw him doing experiments. Why isn't there anything in the notebook from that day?) as for what is included. If this is done at the beginning of a student's career, so that good habits are inculcated at the start and it's clear that high standards will be demanded, the performance of the student will be enhanced and the likelihood of cheating incidents will be greatly decreased. This is also an opportunity for the lab director to make sure that the student puts in the explanations, the plans and reflections, and the summaries that will make it possible to reconstruct what has been done even if the student isn't around. Checks can be made for whether the data have been graphed, so that unusually clean data, as well as experiments that aren't going anywhere, can be identified.

Whistleblowing

Suppose that you are scrupulous in obeying the tenets of scientific ethics but you observe someone else disobeying them. What is your responsibility? Informing someone in authority that misconduct has occurred is called "whistleblowing." While it is important for each of us to challenge scientific fraud when we think we observe it, whistleblowing can be difficult both personally and professionally. The individuals whom we challenge can be angry and vengeful, and institutions are sometimes more interested in covering up a scandal than in letting the truth be known.

It is best initially to discuss the situation with a trusted, experienced colleague or superior. The first thing this person should do is to carefully get your version of the facts, and to examine your understanding and interpretation of the situation. It is possible that you are misconstruing what you have observed, and it is wisest to be sure of your ground before embarking on a formal allegation of misconduct. If possible, the suspect should be given a chance to explain or rectify the situation. If this cannot be done satisfactorily, then it is appropriate to initiate more formal, institutional proceedings.

It is useful to consider the policy of the Office of Research Integrity (ORI) in the Department of Health and Human Services:

> The whistleblower is an essential element in the effort to protect the integrity of PHS supported research because researchers do not call attention to their own misconduct. Prior to making an allegation of research misconduct a whistleblower should carefully study the policy established by the institution for responding to such allegations to determine what information should be included in the allegation, to whom the allegation should be reported, what protections are provided for the whistleblower, and what role the whistleblower will play in the ensuing proceedings. (http://ori.dhhs.gov/misconduct/Guidelines_Whistleblower.shtml)

Because of the possibility of retaliation against the whistleblower, the PHS regulation obligates institutions to protect "to the maximum extent possible, the privacy of those who in good faith report apparent misconduct" and to undertake "diligent efforts to protect the positions and reputations of those persons, who, in good faith make allegations." A good faith allegation is made with the honest belief that scientific misconduct may have occurred. "An allegation is not in good faith if made with reckless disregard for or willful ignorance of facts that would disprove the allegation."

On the role of the whistleblower in misconduct proceedings ORI policy states

> It is the responsibility of the investigative body and ORI, not the whistleblower, to ensure that the allegation is thoroughly and competently investigated to resolution. Therefore, once the allegation is made, the whistleblower assumes the role of a possible witness in any subsequent inquiry, investigation, or hearing. For purposes of the scientific misconduct proceedings, the whistleblower is not the equivalent of a "party" in a private dispute between an "accuser" and "accused."

The big practical problem with whistleblowing is that it may land not just the miscreant, but also the whistleblower and a whole set of institutional officials, in trouble, at least temporarily. Lives are disrupted, personal and institutional reputations are damaged, time and resources are diverted, personal relations become antagonistic. At least those are the worst-case scenarios. The point of doing this, which makes all the pain worthwhile, is to uphold the purity of the enterprise, to deter and punish wrongdoers, to reaffirm values, and to demonstrate to the community that we can wash our own laundry, that principle at least sometimes takes precedence over convenience.

Take-home messages

- Consider the potential applicability of your research, even if at first glance you think of it as purely basic.
- Learn early on how to seek research funding.
- Take your research seriously to ensure its purposiveness, rigor, and objectivity.
- Do your best to apply the scientific method to your research by rigorous testing of a clearly formulated hypothesis.
- Avoid all conflicts of interest—both overt and covert—that might skew your experimental design or data interpretation.
- Avoid research misconduct in any form, including plagiarism, falsification, and fabrication.
- Report potential scientific misconduct to the proper authorities.

References and resources

LaFollette, Marcel C. 1992. *Stealing into Print: Fraud, Plagiarism, and Misconduct in Scientific Publishing*. Berkeley: University of California Press.

National Academy of Engineering, Institute of Medicine, and National Academy of Science. 1995. *On Being a Scientist: Responsible Conduct in Research*. 2nd ed. Washington, DC: National Academy Press.

National Science Foundation. Definition of research misconduct. Federal Register, vol. 67, no. 52 (Monday, March 18, 2002), p. 11937. http://www.nsf.gov/oig/resmisreg.pdf.

U.S. Department of Health and Human Services. 2005. Public Health Service Policies on Research Misconduct; Final Rule. Federal Register, vol. 70, no. 94 (May 17), pp. 28386, 28388. http://ori.dhhs.gov/documents/42_cfr_parts_50_and_93_2005.pdf.

11 KEEPING A NOTEBOOK

The first thing you should do after choosing a research project is to start a notebook. Then use it every day. Start the day by writing down what you intend to do and why. As you perform experiments and make observations, do calculations, and develop ideas, write them down. And end the day by summarizing what you have accomplished, and what needs to be done tomorrow. This takes time, but it's a good investment in terms of keeping your research moving ahead and avoiding waste and delay later.

Why keep a notebook?

Keeping a notebook is one of the most standard and characteristic things that a scientist does; it's almost a caricature of what it means to be a scientist. Yet it's striking how poorly it's often done, both by students and by their mentors. Indeed, once professors stop working in the lab and being the primary collectors and recorders of data, they may stop keeping a notebook altogether, even though many of the functions of notebooks apply to any part of an intellectual life, not just the recording of data.

A research notebook serves numerous important functions:

- It's a record of work done, whether in the lab or in the field. If you know what you did, you won't have to repeat it.
- It's a place for thinking on paper, for recording and exploring ideas, and for noting significant points gleaned from reading and listening to lectures.
- It's a place to record and structure results for use in writing papers and grants. (In drafts of a manuscript—though not in the final version—it may be useful to give notebook page references for results, so that supporting evidence can be readily found later if needed.)
- It's a way to communicate with your research advisor.
- It's a reference for use by others in your research group, or by others who may need to consult and understand your results.
- It's a place to list sources of reagents, dates of purchase, batch numbers, etc. This can help you track down lot-to-lot variability.

- It's a place to list key references, particularly to new methods and reagents.
- It's a place to comment on anything that went wrong during an experiment that led to unexpected results. Many breakthroughs in science have come about as a result of inadvertent mistakes or changes in experimental conditions combined with good record keeping and keen observation.
- It's a permanent record, which makes it possible for questions that arise later (perhaps well after you've left the lab) to be answered. More often than you might think, the research director or another member of the lab will want to refer to your notebook to see where some published data originally came from or how a particular procedure was carried out.
- It's a place to maintain a training record when learning a new technique or instrument. Practical tips and hints and your instructor's answers to questions may be more valuable than the formal material in the instruction manual and should be recorded.

In addition to a research notebook for each member of a lab, a separate notebook should be maintained for each major instrument to record use, calibration, need for repair, etc.

The notebook as motivation

Another important function of the research notebook, one that is not sufficiently recognized, is as a motivator: a reassurance that progress is being made, that interesting things are being found, and that the future looks bright. The physical chemist John R. Platt, in his 1962 book *The Excitement of Science*, has a vivid chapter titled "The Notebook as Motivation." He cites the great physicist Enrico Fermi as his ideal example.

> Fermi would work all day . . . on whatever problem he had chosen, proceeding steadily from order-of-magnitude estimates of the major factors down to fine details. Every particularly pregnant result went into an indexed notebook. Other notebooks held data, and one was a "Memory" of useful equations and numbers. . . . [His] attitude in his work was that it was not worth doing anything unless you gave it your whole attention. . . . It was no use to master something unless you made a permanent record of it in the notebook; and no use to make a record unless it was something you understood. (135–36)

Platt goes on to generalize about the important but often neglected role of the notebook as a motivating device. He points out that the permanently bound notebook facilitates formal reasoning by emphasizing continuity of effort: "The powers of civilized men are not due to an increase in our thinking ability but to a decrease in our loss rate" (138). The notebook is a stimulus to creativity by giving the mind a record of what it has accomplished, a reminder of what needs to be criticized, an the anticipation of new accomplishment. It distances in time creativity and criticism. It requires focus on one problem at a time, overcoming tendencies to wander. It urges you to make a new contribution every day.

How should a notebook be arranged?

The definitive book on keeping a notebook is *Writing the Laboratory Notebook* by Howard M. Kanare (1985). Much of what follows is adopted from this book, which codifies and elaborates on well-established practices. As Kanare spells out in his chapter 5, every notebook should have certain characteristics:

- An exterior title, so it can be identified on the shelf. This can be a project name or abbreviation, but most commonly is a sequential code (for example, the researcher's initials plus a number).
- Numbered and dated pages. If the pages are not prenumbered, do it yourself by hand, enclosing each number in a circle so it won't be mistaken for a date or data.
- Entries written in indelible, nonfading black ink, preferably ballpoint, which is resistant to liquid spills. Entries should *never* be in pencil.

The first pages should tell future readers—including yourself—what's in the rest of the book. Include the following:

- On the first page, your name, department, organization, and the dates of the first and last entries.
- A table of contents, which should be filled in as the notebook is written. Leave about one line per notebook page. Each entry should include the date, a brief, readily comprehensible description of the subject, and page numbers.
- A preface or introduction. Though often neglected, an introduction is useful in orienting you and others who may consult the notebook

about the purpose and context of the work being recorded. Include such things as the goal of the research and its status when the notebook was started, the names of your supervisor and coworkers, the sponsoring agency with title and grant number, and perhaps a statement of where the work is being done and where other records and samples are stored if the research is being carried out at a variety of sites.

- A list of abbreviations, symbols, codes, and other shorthand that will be used throughout the notebook. Don't forget to update this page as work proceeds.

After these important preliminaries, the body of the notebook will record your day-to-day plans, results, and ideas. If an entry or sequence of entries jumps between noncontiguous pages, note "Continued on page ____" at the end of the page preceding the break and "Continued from page ____" at the top of the new page. If the gap corresponds to an interruption of the work, indicate at what stage, and for how long, the work was interrupted.

It is almost inevitable that occasionally an idea or some data will be scribbled on an odd piece of paper rather than entered properly in the notebook. If so, this material should be pasted into the notebook or copied as soon as possible. Particularly if the material is potentially patentable, pasted loose pieces of paper should be signed, dated, and witnessed.

What should your notebook contain?

One of the most common flaws of research notebooks is that they contain only bare, poorly annotated tables of numbers, with hardly any commentary or context. This may suffice at the moment the data are recorded, but it will be hard for you or anyone else to go back to these pages later and understand what was done and why, and what conclusions might be drawn from the results. A notebook should be a comprehensive record, with enough detail that another scientist competent in the area could repeat your work. It should be a place where information from the literature is written down and integrated into the ongoing work. And it should be a tool for thought: a place for recording hunches, ideas, and odd thoughts that might stimulate future work.

A useful way to proceed is to envision each day's work, or each subproject within the larger project, as a small manuscript. Each subproject might lead to a figure, a table, or some mathematical, computational, or statistical result within a larger manuscript. Viewing the successive pieces of your project in this way will help you focus on achieving individual results that will contribute to the whole, and will give you a sense of steady progress.

Start as if you were writing an article for a scientific journal, with a descriptive title and an introduction outlining the purpose, background, and plan of the work. Indicate why you expect your approach to contribute to progress on the overall project.

Follow that with a materials and methods section, where your approach to the problem is elaborated, listing equipment, reagents, and analytical and statistical approaches. These might refer by page number to material developed previously in the notebook, rather than being spelled out from scratch each time. But be careful to note all of the details specific to this particular experiment: instrument settings and calibrations, reagent batch numbers, environmental variables such as room temperature, and so on.

Next comes a results section, where you actually perform the experiments or calculations and record the outcome. Preliminary preparation of a table in which to record data is usually advisable; it will remind you of all the relevant variables and help avoid omissions. Record data or observations immediately, as you obtain them, writing clearly what you did and what you saw. Be neat, so that you and others can decipher what you've written. If something unusual happens that might affect the results, write it down. If you make a mistake in recording, draw a single line through the entry, and write a brief explanation of the error. If results are generated by computer, single pages can be pasted into the notebook. If the material is more voluminous, make a clear reference to where it is stored and cross-reference the computer file or printout to the notebook page.

After the day's results are obtained, it's time to analyze them in a discussion section. This might involve graphing, recalculation (e.g., subtraction of baselines, normalization, multiplication by calibration factors), statistical analysis, ideas, and speculations. Be sure that graphs are clearly referenced to the original data and are appropriately labeled, with units, symbols defined for each data series, and, if possible, error bars. Write down how the results accord with your hypothesis. Do they suggest the need for further work, for new experiments or calculations, or for changes in the experimental plan?

Finally, you should write a brief conclusion to the day's work, along the lines of an abstract for a journal article. What was your goal? What did you do? What did you find? Was the experimental design effective, or does it need tweaking? Was your hypothesis supported by the results? What next steps seem in order?

Schedule some time at the beginning of each day to write the introduction and methods sections for each day's work, and some time at the end of the day to write the discussion and conclusions. This may seem hard to achieve, given that most of the day will be devoted to collecting results, but the time saved in thoughtful setup and analysis leading to more efficient next steps and a better understanding of the evolving project will more than repay the investment. "A stitch in time saves nine" is pertinent here.

Using the notebook as an explicit place for planning, thinking, and reflection is very important, yet too rarely done in our experience. Relate each experiment and entry to what has gone before and to the general chain of reasoning that motivates the whole project. This will be particularly valuable when someone—either you or someone else, such as your research director—looks at the notebook later and tries to figure out just what you were trying to accomplish by doing this particular experiment. Writing about it might also give you ideas about better ways to do the experiment, or lead you to realize that, as conceived, it is no longer pertinent.

Of course, it's hard to know whether to stay on course or change direction unless you've analyzed your data and seen what direction it's leading you. Too many students collect data for months before summarizing and graphing or tabulating them to look for trends and agreement or disagreement with existing ideas. This is a big mistake. You should always analyze and examine your data as soon as you can. Then you can tell whether an instrument is out of calibration or a key reagent has gone bad, or whether there is some fundamental disagreement with expectations that may betoken a truly exciting scientific discovery.

The notebook as patent documentation

The practices listed in the previous sections are important in maintaining a patent-worthy notebook. Write legibly, use indelible ink, use a bound notebook with numbered pages, record entries consecutively, don't remove pages, paste or tape loose sheets into the notebook (then date and sign each insert, with your signature flowing across insert and notebook page), explain abbreviations or acronyms, cross out incorrect data and add explanations of why this was done. Record your ideas, not just experimental results. Make the record complete and self-explanatory by structuring it as a mini-paper.

A couple of things are different, however, if there's potential for a patent. For one thing, don't write that a result was "obvious" or that a line of experimentation will be "abandoned," since "obvious" and "abandoned" have specific legal meaning in patent law.

According to the best practice for patent purposes, each page of the notebook should be signed, dated, and witnessed by you (the person who did the work) and by someone who is not a coinventor but who has the technical qualifications to understand what has been done. You should sign "Recorded by [name] on [date]," and the witness should sign "Witnessed and understood by [name] on [date]." This signing should be done as soon as convenient after the end of the experiment, most likely at the end of each day. While this is generally the practice in an industrial laboratory, it is relatively uncommon in academia. The absence of such documentation is not an absolute bar to obtaining patent protection; it just makes it harder if your claim is challenged.

Since in patent work the notebook—preferably witnessed and signed each day—is a key piece of evidence of priority, there may be concerns about putting down some of the less formal and objective aspects of your work—hunches or guesses or very preliminary interpretations of your data. Do you really want to put down things that are not well substantiated? Or, perhaps more importantly, not followed up promptly, since an important part of proving a patent claim is that you have diligently and consistently pursued the idea since you conceived it. An idea jotted down a year ago and only recently picked up again may not meet this test. If you're working in a company where notebooks are crucial to patent filings, you'll be instructed in these details. If you're a student or postdoc working in a university lab, it's best to err on the side of writing everything down.

Electronic notebooks

Should you keep an electronic notebook on your computer, instead of a bound paper notebook? An increasing number of research groups, in both academia and industry, are doing it. The increasing availability and decreasing cost of laptop, notebook, and tablet computers make them nearly as convenient as bound paper notebooks for data recording at the lab bench and in the field. And wireless networks make it possible to send or receive data and to access online resources quickly and efficiently. Electronic notebooks can provide a number of significant benefits; some are simply implemented in word processing or spreadsheet programs, while others require programming or hyperlinking.

- Searching and indexing can be done quickly and automatically. Referring to previous work, which may have been forgotten or hard to find, and therefore may be needlessly redone, becomes easier.
- Data need not be copied, rekeyed, or pasted in.
- The full original data set can be maintained in the notebook.
- Data in various formats—numerical tables, photos, graphs, sketches, videos—can all be accommodated and consolidated.
- Hyperlinks to other information (standard tables, references, etc.) can be established.
- When working with computerized instruments, the settings of parameters can be entered automatically or in response to prompts. Reminders of periodic calibrations and maintenance can be issued.
- Standard data tables can be constructed automatically using templates.
- Notebooks can be backed up regularly so that physical loss or damage is no longer a danger.

Electronic notebooks possess additional virtues in collaborative research, if suitable data communication and storage protocols are implemented.

- Data can be shared and common notebooks inspected at a distance (assuming suitable read/write permissions).
- Notebooks kept on a central server can be backed up regularly, with a backup maintained off-site to guard against data loss.
- Both student and supervisor can have ready access to the notebook.
- When working with major shared instruments, scheduling can be accomplished through the same computer interface, and notes can be left about instrument condition. This can increasingly be done at a distance via Internet-controlled interfaces.

The basic problem with using an electronic instead of a paper notebook is the ease of altering computer entries. Alteration is of concern both in its potential for research misconduct (fraud or falsification) and in making uncertain the dates of invention or discovery for patent purposes. Handwritten notebook entries are basically permanent, since if they are recorded in ink then any changes leave physical marks on the paper. By contrast, changing a computer entry can be no more difficult than changing the file and rewriting it. You realize that you need to do it a few days later, and want to maintain the original date? No problem; just temporarily reset the date and time on your computer.

Even without dishonest intentions, there are problems. Entries in a paper notebook are individually dated, while an electronic notebook typically records only the date a file was last modified. If you keep your research notebook in a single computer file, then its modification date will be changed each time you save an entry. To register the correct date, you would have to start a new computer file each day. This is not necessarily a bad idea, but it undermines the idea of having everything in one searchable file. Still, there are technical solutions to this problem. There are codes that can be embedded in each file to validate the dates, and data can be sequentially recorded on write-once CDs or DVDs or in an off-site database (perhaps maintained by a third party for verification purposes). In fact, requirements for signing and dating can be met more readily, assuming the electronic versions meet legal standards. One can expect further advances along this line, as well as clarification of the legal status of notebooks kept on computers. Some large companies have gone this route, and they're certainly concerned about verification and patent issues; so it can be done, but the software and the implementation mechanisms are specialized.

Even if your research advisor requires the use of bound paper notebooks, it will make sense to keep some information on the computer. Maintaining a table of contents and an index, for example, may be more easily done on a PC. Log each experiment or notebook entry in a database or spreadsheet program, noting keywords as well as date, page, and topic. You can then generate an index by finding keywords in either the topic or keyword fields. Whether or not you keep these records on a computer, however, at least the table of contents should also be written in the notebook itself.

Research protocols and documentation of methods may be maintained on a central server and thus made available to all members of the research group. This can have great advantages if one member of the lab develops improvements over what has been done before. Increasingly, too, data are computer-generated and so voluminous that computer storage is the only sensible way to proceed. Keeping the data on a central server, with frequent off-site backups, is increasingly the practice of large labs in the corporate sector. These systems have well-validated security and date-stamping mechanisms. With such mechanisms in place, patenting concerns and the possibility of fraudulent data manipulation are minimized. In the more casual academic environment, intellectual property may (and probably should) be of less concern. But even in universities there is an increasing drive to protect and capitalize on intellectual property, and data integrity is always an issue. Whether electronic notebooks help or hinder the reliable and efficient conduct of science is an evolving topic.

Who owns the notebook?

In the vast majority of cases, funding agencies (whether governmental or private) officially make their research grants to institutions such as universities, not directly to faculty researchers. The agency and the institution, of course, expect the faculty member and their research group to carry out the work. But if something goes wrong, if a question is raised about data integrity or mismanagement of research funds, the agency will hold the institution legally responsible, since it was the institution that received the money to carry out the research. In extreme cases, the agency may demand return of the funds, which—with the inevitable legal and administrative expenditures that attend such situations—can cost a university hundreds of thousands, even millions, of dollars.

The institution, while providing oversight of financial matters, understandably and properly relies on the principal investigator (PI) of the grant, usually the faculty researcher, to exercise prudent fiscal management and (especially relevant to this chapter) to assure the reliability of the research results. This means, among other things, that the lab notebooks of all the researchers in the lab—grad students, postdocs, visiting scientists, technicians, undergraduates, and the PI—are the property of the university and must remain with the laboratory. They should be maintained there until the lab is closed and the PI retires.

PIs may allow students to make copies of the notebooks; indeed, this is prudent backup procedure as well as a way of allowing students continued access to their own work once they've left the lab. But the original notebooks must stay in the lab and must be accessible to the PI at any time. Advisors may, being polite, request that students show them their notebooks—but unannounced inspections are in order at any time.

Take-home messages

- Use your notebook to record details of experimental work—the results obtained and their potential interpretation—and to put down your thoughts regarding future experiments and research directions.
- Include all pertinent details of how an experiment was done, and note any anomalies. Your description should enable someone else to replicate your experimental conditions.

- Do not put off describing an experiment to the following day—many important details will be forgotten.
- Keep careful records in your notebook, since it might someday serve as a legal document in support of a patent application.
- Make a copy of your notebook, as the original is the property of your advisor acting as agent of your university.

References and resources

Kanare, Howard. 1985. *Writing the Laboratory Notebook*. Washington, DC: American Chemical Society.

Oak Ridge National Laboratory. Electronic Notebook Project. http://www.epm.ornl.gov/~geist/java/applets/enote/.

Platt, John R. 1962. *The Excitement of Science*. Pp. 134–39, "The Notebook as Motivation." Boston: Houghton Mifflin, Boston.

12 WORKING WITH OTHERS

The image of the lonely scientist, working in his solitary lab till all hours of the night, is ingrained in popular culture. And indeed, your graduate and postdoctoral research may entail long hours, many of them spent alone in the lab or library. Nonetheless, you're also part of a larger social fabric. An important part of the quality and type of science that you do, and the success you have in your career, will be dependent on your relations with the other people who impact your research. Most obvious are your advisor and your labmates—the other students, postdocs, and technicians in your research group. Your quality of life during working hours will depend on them, and you can learn from their different approaches to problem solving and critical thinking.

Your advisor

We've already talked about choosing your research advisor, and what you should expect from her or him. If you're lucky, your advisor will be attentive, careful of your professional development, a good mentor, and sensitive to those periods when you need special attention. Some advisors are like that, but many aren't, and you'll have to figure out how to work with what you've got. It will be particularly important to make sure that you are assigned an appropriate project and that significant parts of the project are not inadvertently (or worse, purposely) given to another worker in the lab, setting up a competition. Advisors should be reasonably available to offer advice and guidance at crucial points. They should be willing to let students move on when their PhD work is over, rather than milking their work for additional publications or service. As a student, however, you should take the lead in seeking career advice from your advisor and working with him or her to create an individual development plan.

If your advisor isn't readily available for drop-in discussions, make an appointment. Make it for a defined length of time, and indicate the specific things you want to talk about. If you come prepared with your own agenda, you are much more likely to get results that will be useful to you than if you just wander in and say "I want to talk." Respect your advisor's busy schedule when asking for advice.

Some advisors, on the other hand, are always wandering through the lab and looking over students' shoulders. If you are being asked for hourly or daily progress reports, you might be grateful that you have the sort of advisor that is really involved with the work of the lab. But you might also find it creates too much pressure. If that's the case, either ask to talk at a later, more convenient time, or use the frequent visits to prime your own efforts to analyze and think about the results of your experiments and what comes next. If the pressure becomes really uncomfortable, ask for a private meeting and tell your advisor that you really value their attention and input but think you'd perform better if you had to report less frequently. Most advisors will respect your request.

An occasional advisor is so busy, or so involved with personal difficulties, that he or she rarely asks about data or the status of projects. If such disengagement persists, you and your labmates would be justified in bringing this behavior to the attention of the department chair or director of graduate studies. Be sure to do so in a way that indicates concern rather than antagonism.

You should not expect your advisor to be a friend or personal advisor (though that may happen), and you should definitely avoid more intimate relations, at least until the mentor-mentee relationship is ended. Relations should be on a professional level, and since you are an adult you should be able to do your part in seeing that they are.

Your labmates

If you're in a discipline that emphasizes fieldwork (e.g., geology or ecology), you may define and carry out your research relatively independently. Otherwise, you're likely to be part of a group of students and postdocs working in close proximity. Your interactions with these labmates will be more frequent and on a more equal footing than those involving your advisor. Your labmates will provide acculturation when you're new and companionship throughout your time in the research group. Some may become part of your social circle. They'll give scientific and personal advice, loans of equipment and special reagents, assistance with troubleshooting technical problems, and constructive criticism at lab meetings and when you're preparing a talk or poster presentation. They'll teach you about the special computer programs and lab protocols used in the group. They may collaborate and coauthor with you. They'll provide support at critical times: when you're having personal difficulties, or when you're coming up for a crucial exam. You should seek their feedback on manuscripts and grant or fellowship proposals. And you should become familiar with their projects and findings. This could result in future collaborations, even after you all leave your advisor's lab. It could also provide you with ideas for new research projects—but be careful not to appropriate theirs.

Be very wary of romantic relationships in the lab. What initially seems idyllic can go bad and produce great awkwardness for both parties, as well as the people around them. Especially avoid such relations with a supervisor. If you do get involved with a peer, keep the lab relationship entirely professional.

The kinds of difficulties that can arise with your labmates—aside from romances gone sour—include annoying personal habits, encroachment on space, misuse of common equipment or reagents, disputes about who is working on what aspects of a project, and quarrels about collaboration and joint authorship. As in other contexts, the best way to handle problems is to be calm and objective, but let your concerns be known. If a labmate plays music that annoys you, ask them nicely to turn down the volume or use earphones. If someone encroaches on your portion of the lab bench, point out politely that you've been assigned this area, which is necessary for your work, and ask them to keep out. You may need to work with the encroacher to devise a mutually satisfactory solution, perhaps finding shelf or refrigerator space elsewhere and rearranging things, or you may need to ask the advisor for help in devising a solution.

If a labmate is misusing common equipment or supplies, then again pointing out politely but firmly that there is a problem is the first step. There should be agreed-upon rules about who in the lab is responsible for cleaning and maintaining the various pieces of common equipment. If there is not, then group members should suggest to the advisor at a lab meeting that such allocation of responsibilities is needed, or you may work it out among yourselves. (If you do so, tell the advisor, in case they decide to make the assignments themselves, which could lead to embarrassment.) Similarly, if there are other supplies or reagents that several people in the lab use, decide on a suitable apportionment of responsibilities. If somebody screws up, there's then a regular procedure that can be understood and appealed to. Anyone can make a mistake once, so don't be too harsh; but if it persists, you may need to express concern to the advisor, which could lead to restrictions on the miscreant. Such situations should be rare. Most people learn to perform well, or at least adequately, if properly trained and subjected to peer pressure.

Collaborations outside the group

Working with scientists outside your research group can be productive and rewarding. Outsiders possess a different range of knowledge and set of skills, and they may provide an analytical technique, a theoretical insight, or a unique reagent that your group doesn't have. They may also offer a useful perspective on your ideas that you would not find within your own group, given its relatively narrow focus. You should definitely try to make contact with others, to broaden your horizons and to get the reactions from the wider world to which you hope your science will appeal.

Before you embark on a collaboration with others, however, you should check with your advisor, who may be wary of sharing unpublished data, credit, or research plans. Advisors vary widely in their attitudes toward such things, and while we tend toward the free-communication end of the spectrum, your advisor is the person you need to pay attention to.

Collaborations with other groups can often lead to publications, and here the question of who should be a coauthor and who just thanked in the acknowledgments section raises its (not necessarily ugly) head. As usual, the best thing is to be up front about your concerns, to discuss at the outset—advisor to advisor as well as student to student—whether this collaboration should be expected to lead to a joint paper, or what level of involvement would make that possible. A memo or e-mail putting the understanding in writing would be prudent. See chapter 21 for a more thorough discussion.

Credit, priority, and sharing

Scientists not only want to understand nature, they want to be the first to do so. The first to publish gets the credit. Priority in discovery is important for recognition by scientific peers and perhaps the public at large, for funding of future research, and for psychological motivation. The competitive urge in successful scientists is just as highly developed as it is in people who succeed in other walks of life. But *excessive* competitiveness is even less appropriate in science than elsewhere. Science is built on openness, on cooperation, and on trust. Don't be overly secretive; sharing information is usually more productive than withholding it.

A basic tenet of scientific work is to be sure that your data are verifiable and reproducible, both by yourself and others. A corollary of this "reproducibility" principle is that experimental materials that may not be commonly available (e.g., microbial strains, immune cells and reagents, cell cultures, or knock-out mice that you may have developed) should be made available to others so they can check your work. Likewise, large collections of data (e.g., atomic coordinates, and the x-ray intensities from which they are derived, in macromolecular crystallography) should be made available to others who wish to check the correctness of your conclusions.

It is a remarkable testimony to the cooperative spirit in science that such sharing is accepted as standard practice. Nevertheless, in these highly competitive times, sharing occasionally breaks down, and the ethical issues involved are not always simple to resolve. For example, a competing laboratory may want your special mutant not so much to check your results as to conduct its own experiments. All well and good, but suppose you also had the idea of doing those experiments, and the other lab is bigger and can move faster than you can. Are you justified in withholding the mutant? If so, for how long?

Or suppose that, as a crystallographer, you have invested several years in acquiring your data and publishing an initial report, and now want time to thoroughly analyze and explore the implications of your results. The standards of science (and of many crystallographers) state that the data should be published or deposited in a publicly accessible computer database. Are you justified in withholding the data?

An additional complexity of modern science, particularly it seems in biology, is that discoveries can be turned to commercial profit by the biotechnology and pharmaceutical industries. While this may be a good thing for society, it means that the sharing of materials or data by a researcher may result in loss of the financial benefits stemming from the research. University rules may require that potentially patentable materials be subject to "materials transfer agreements," or MTAs. For example, see the MTA from the National Heart, Lung, and Blood Institute at NIH, http://www.nhlb.nih.gov/tt/docs/sla_mta.pdf.

The decision about whether and under what circumstances you should share material or data should be made primarily by your advisor, since he or she is the one who bears legal responsibility for the work and it is to the lab that the material belongs.

Despite these complexities, the general principle is that sharing and openness is the norm. Not only does it foster reliability and faster progress, it also invites reciprocity. If you share with someone, they are more likely to share with you. For further discussion, see the articles "Data Sharing" and "The Culture of Credit," listed under "References and resources" at the end of this chapter.

Departmental staff and others

Who are some of the other people with whom you might work or come in contact? Think of the staff in your department: the receptionist, the secretaries and clerical help, the person who delivers the mail, the stockroom attendant, the Internet staffer. These are all people who make your daily life run much more smoothly than would otherwise be the case. Appreciate their efforts and be nice to them. Say hello in the morning and when you pass in the hall. It will make both your day and theirs more pleasant, and you can never tell when a nice word about you to some third party may be helpful.

It's too bad that the office staff in our departments don't know more about what we're doing in the lab. They are citizens, after all; they vote and influence their representatives and their friends, who are also voters. Especially if you're at a public institution that depends to some extent on government support, it's worthwhile to encourage goodwill among members of the public. That will be a lot easier if the receptionist goes home each day thinking how interesting and pleasant the job is than if she or he leaves bored or angry.

You should get to know the department chair, the director of graduate studies, and the other professors in your department and related departments who do work relevant to your research interests or teach courses that you must take to satisfy the requirements for your degree. These are the people who will grade you, sit on your oral exam and thesis committees, and serve as references when you look for jobs in the future. They may well also have the expertise that will give you the needed clue to break through a logjam in your research. Faculty may seem busy and unapproachable; in fact, they are busy, but most welcome the chance to talk with a student or postdoc who is genuinely interested and intelligent. They don't want people to butter them up, but they enjoy teaching—that's why they're in academia, after all—and working one-on-one with a student who has an interesting research question is particularly rewarding. Intelligent questions during and after class will also make a good impression, as well as deepening your own knowledge of the material.

Most departments have social gatherings at holidays and receptions for prominent visiting speakers, as well as less formal get-togethers before or after seminars and perhaps at the end of the week. These are great opportunities both to network and to relax with friends.

Who owns a project?

As research groups get larger and collaborations become more common, confusion and disputes can arise about who "owns" a project: Who should be listed as authors on papers arising from the research? Who will determine the future direction of the project? Who should have control over special reagents? As a graduate student you need some portion of the project that is unquestionably your own for your thesis. If you're a postdoc you need to establish your independent contribution to get a good job. Recognition of your contributions to an important piece of work is crucial for career advancement, for grant or fellowship funding, and for your own psychological motivation.

Disputes over who owns a project are not uncommon and can turn nasty unless dealt with early and unambiguously. Such conflicts often arise when the advisor either inadvertently assigns the same or closely related projects to more than one person, or deliberately sets up a competition so that the lab will progress faster (and the advisor's career will prosper) regardless of what the effects on students. They can also arise when someone, from your lab or another, is enlisted to help with a small part of the project but then blends it into work they are already doing so that it seems to be part of their project rather than the project they are assisting.

How do you reconcile your need to develop an independent scientific reputation, to get clear credit for the contributions you have made, with the equally valid need to draw on the expertise of others to get the job done? How can you avoid losing appropriate credit for your work in a welter of little collaborations?

The best way to deal with these issues is to be up-front and explicit at the beginning of the relationship or collaboration. "I'd appreciate it if you would help me out with this. I think it should be just a couple of hours once we get the system running, and I'd be happy to list you in the acknowledgments of my paper. If this really turns into a major collaboration, with your spending several days or weeks and gathering results that form a significant part of the content of the paper—at least one figure or table—then maybe you should become a coauthor. But at this stage I don't think I'll need that much of your time or effort." If the help is limited in this way, acknowledgment rather than coauthorship is the appropriate mechanism of recognition. After all, being an author implies being intellectually responsible for everything in the paper (see chapter 24)—a responsibility that someone who has just helped you with a small part of the project should not want to accept.

The advisor can be very helpful in defining the boundaries, by making it clear to the rest of the group just what it is you're doing and how it relates to the other projects. When a new person joins the group, the advisor should introduce them at a group meeting and say, "This is Mary Smith. She's just joined the group, and she'll be working on solvent effects on DNA condensation. This is something we've wanted to do for some time. It will complement John's studies on ionic structure effects and Dmitri's on the theory of dielectric constant behavior. Mary, you'll be using the same light scattering equipment as John, so, John, I'd appreciate it if you'd show Mary how to use it and how our analysis software works. And your results will be very useful as Dmitri develops his theories; I hope that the two of you can talk and keep in touch with each other's work, since Dmitri may be able to suggest experiments that would help test those theories. There may be possibilities for coauthorship in these interactions, but Mary will be first author on papers dealing with solvent effects. I hope that you'll come to me so we can talk over possibilities for coauthorship if things seem to be moving in that direction."

A memo laying out the policy on collaboration, cooperation, and authorship within one of the authors' research groups, and in its dealings with other researchers, is given below.

> Our research, with its wide variety of techniques, theories, and ideas, often requires interaction with others—both within and outside our research group. At the same time, our focus on a limited set of research problems means that there will be overlap of interests and efforts. Complex issues of intellectual ownership can arise as a result. My intention here is to define what I think are some reasonable policies and rules.
>
> Unless otherwise agreed, each project that seems likely to lead to a publishable paper will be identified with a single worker in the lab—

a graduate student, postdoc, scientist, or visitor. (Undergraduates and lower-level technicians will generally be assisting someone else. Rotation students are to be considered on a case-by-case basis.) The worker will have the prime responsibility for preparing material, collecting and analyzing data, developing theoretical explanations, and writing a first draft of the paper. Each such paper will be coauthored by the worker and the primary investigator (PI), unless the PI decides that he has contributed so little, or understands the work so poorly, that he chooses not to be an author.

This policy is based on several principles. The first is that students and postdocs are at a learning stage of their career, and learning the major experimental and theoretical techniques used in their theses and postdoctoral projects is an important part of their education. They should not rely on others to do major portions of their work.

A second principle is that students and postdocs, who will be going on to independent careers, need publications on which they are clearly the principal author. The coauthorship of the major professor is a generally understood adjunct, based on their suggestion of a problem area and broad approach and their provision of guidance and resources, as well as on specific intellectual input.

At the same time, it is understood that the lab is a cooperative enterprise, in which all the workers share an interest in common scientific problems and are willing to contribute their special expertise to helping others solve their problems. For example, everyone participates in group meetings and should be active in providing constructive criticisms and suggestions. And everyone who uses a common technique (DNA preps, light scattering calibration, etc.) should contribute to a continually updated lab methods book.

When it comes to specific research projects, there are different levels of sharing, which call for different types of acknowledgment and involvement in the published work. The first is the most casual: calling attention to a relevant paper, a helpful hint about a technique that isn't working right, etc. This is basic neighborliness, and needs no special acknowledgment.

The second level of sharing involves a substantial but limited commitment of time and effort. Examples are spending several hours or days teaching someone a technique or preparative method, providing a DNA sample that took a lot of effort to prepare, performing a nonroutine analysis on a one-shot basis, or providing access to a specially written computer program. Such efforts deserve an acknowledgment at the end of the paper. Within our group, everyone should be willing to make such an effort to help another. You may hope for, but should not expect as your right, such help from people in other labs.

The third level of sharing is one that involves enough continuing effort as to merit coauthorship. Each worker should contribute a specific expertise, generally over a prolonged period of time, that leads to a substantial portion of the final paper. All coauthors should understand the final manuscript, be able to defend its conclusions, and take responsibility for it. If it has not been agreed at the beginning of the project that there will be coauthors, then the worker who began the project must agree to the coauthorship. By the same token, the original worker should not ask for, or accept, prolonged assistance and collaboration unless they are willing to grant coauthorship. Both parties (and the PI) should make explicit agreements about coauthorship before substantial collaboration begins. As noted in the third paragraph, however, it is generally desirable, particularly for graduate students, to learn and carry out themselves all the major aspects of their thesis research.

Careful and up-front collaborative and coauthorship arrangements are equally important when you work with someone from another lab. It should be standard practice, at the point when any major commitment of effort is being discussed, to discuss who will be coauthors on the resulting paper(s), if any, and who should be satisfied with an acknowledgment.

To begin to implement these policies in our lab, I ask each of you to give me, by the end of next week, a brief description of each project on which you are now engaged or on which you expect to begin shortly. (Each project, to reiterate, should correspond to one anticipated paper.) Your description should include the goal of the project (i.e., the question that you hope the paper will answer), the approaches you will use, and whether you have arranged a collaboration or coauthorship with someone. In case more than one person claims essentially the same project, without an agreed collaboration, the PI will make the assignment after discussion with the concerned parties. Likewise, in the future, whenever you embark on a new project, you should give the PI this information about it. Not only will this clarify your intellectual property rights, it will contribute to a better initial focus of your research plans.

Can you take the project with you?

Equally difficult, especially for postdocs who are expecting to move on to independent research positions, are questions about who owns the project's main direction. What if it was initially your advisor's idea? What if it was your own idea? How does this influence your right to take the project with you and pursue it independently? How does this play out in grant proposals submitted by either you or your advisor once you leave the lab in which you are training? These conflicts are not uncommon, and they can be painful.

Our advice, as with so many of these ethical puzzles, is to try to address and clarify the situation ahead of time. When you are applying for a postdoctoral position, one of the questions you should ask concerns the attitude of your potential advisor toward postdocs' taking part of their project to their next job. If you have a choice of postdoctoral opportunities, you would do well to choose one that allows you to do so, since having a project already up and running will give you an important head start, including preliminary data to use in a grant application.

Troubleshooting problem relationships

Even if it's not over such a serious matter as an infraction that requires blowing the whistle, your relationships with your advisor, labmates, or others can go bad. In such cases, it's best not to let bad feelings fester but, instead, to get things out in the open—in a civilized, collegial way. Formal complaints can burn bridges; informal discussions are better if possible. Talking things out can be very helpful, allowing both parties to see alternatives and remove rough edges.

If a straightforward conversation doesn't resolve matters, there are generally institutional resources that can help you find solutions to interpersonal problems. These may include your department chair or director of graduate studies, a departmental committee, or a university-wide ombudsman, conflict resolution office, or grievance office. These university offices are especially experienced and skilled at keeping conflicts low-key. Consider, for example, this description of the modus operandi of an ombudsman:

> An ombudsman is a designated neutral, confidential, nonjudgmental, independent person that can help you figure out how to deal with a problem. An ombudsman will not take sides or tell you what to do. Ombudsmen

usually don't take notes or keep records on your visit and don't testify at hearings concerning a dispute. An ombudsman will listen to you, ask a lot of questions, talk out the problem, give feedback, brainstorm what might be done next, and consider the consequences of various courses of action. If necessary, an ombudsman can act as a mediator or perform shuttle diplomacy to help resolve a dispute. (Agrawal 2001)

Because of their intermediate status, postdocs may not have access to either student or employee grievance mechanisms, though that may be changing as more universities establish postdoc offices. However, they should still be able to obtain advice within their departments or avail themselves of ombudsman or conflict resolution offices.

Develop a leadership style

If you are to develop a successful career, whether in academia or elsewhere, you need to develop your personal and leadership skills as well as your scientific ones. Working effectively with others, and building a strong network of mentors and contacts, is part of what's required. But you also should try to develop a style that enables others to view you as a leader within your research group and department. Sheila Wellington and Betty Spence, in their book *Be Your Own Mentor*, offer this advice on developing a successful style:

- Make others comfortable.
- Get past other people's assumptions.
- Radiate confidence.
- Learn the art of the humorous comeback.
- Be seen as a team player.
- Focus on producing results for the organization.
- Pick your battles carefully.

Being viewed as a positive leader in the group, but not as one who tries to outshine everyone else, will put you in good stead when it comes to recommendations for the next stage of your career.

For the mentor

Managing your students and postdocs effectively is an important part of your success, as well as theirs. Providing motivation and advice, resolving conflicts, and guiding people to the next stages of their careers is a key part of your responsibility. Austin (2003) offers some succinct advice on how to do it effectively:

- Measure outcomes, not inputs.
- Don't measure outcomes overzealously.
- Keep your commitments.
- Treat people decently.
- Remain approachable and listen.
- Establish reasonable policies, communicate them clearly, and stick to them.
- Keep some slack in the daily schedule.

Take-home messages
- Approach your advisor for help whenever questions or problems arise.
- Provide your advisor with periodic reports on your progress. You might have to initiate this exchange if your advisor is busy.
- Be sensitive to your advisor's time commitments and preferences. Find out if he or she welcomes walk-ins or prefers that you make appointments.
- Think of the members of your advisor's research group as your community, a team of coadvisors.
- Observe how your labmates handle different situations and approach problem solving.
- Use the lab environment as a training ground for the fine art of conflict resolution.
- Maintain open communication with your advisor and labmates regarding your roles and responsibilities.
- Discuss with your advisor guidelines of authorship when you are involved in a collaborative project.

References and resources
Agrawal, Alka. 2001. "Butting Heads: Conflict Resolution for Postdocs, Part II." http://sciencecareers.sciencemag.org/career_development/ previous_issues/articles/0980/butting_head_conflict_resolution_ for_postdocs_part_ii.

Austin, Jim. 2003. "Managing Knowledge Workers: Rules for Absolute Beginners." http://sciencecareers.sciencemag.org/

career_development/previous_issues/articles/2380/managing_
knowledge_workers_rules_for_absolute_beginners.

Cohen, Carl M. and Suzanne L. Cohen. 2005. *Lab Dynamics:
Lab Management Skills for Scientists.* Cold Spring Harbor Lab Press.
New York.

Cohen, Jon. 1995. "The Culture of Credit." *Science* 268 (23 June):
1706–18.

Marshall, E. 1990. "Data Sharing: A Declining Ethic?" *Science* 248
(25 May): 952–57.

Wellington, S., and B. Spence. 2001. *Be Your Own Mentor: Strategies from
Top Women on the Secrets of Success.* New York: Random House.

CREATIVITY AND PROBLEM SOLVING

Research, creativity, and problem solving

The essence of research is creativity, yet probably few graduate students or postdocs (or their research advisors) would instinctively use the word "creativity" when asked what is required to do the research that will lead to a PhD and a successful research career. This is not because creativity is not valued but because it's taken for granted.

We think of research as pushing back frontiers, opening up new territory, and of creativity as making something new or solving a problem in an ingenious way. So research and creativity are similar, if not synonymous.

Mihaly Czikszentmihalyi, in *Creativity: Flow and the Psychology of Discovery and Invention*, says, "Creativity is any act, idea, or product that changes an existing domain, or that transforms an existing domain into a new one. And the definition of a creative person is: someone whose thoughts or actions change a domain, or establish a new domain. It is important to remember, however, that a domain cannot be changed without the explicit or implicit consent of a field responsible for it" (28). This definition emphasizes that not only must you do the creative work, you also have to inform and persuade others within your field. We'll talk about communication in later chapters; but for now, remember that a discovery that is not published and talked about does not affect the course of science.

How about more modest, day-to-day problem solving? You're working away on your project, you run into an obstacle, you need to solve it, what do you do? The first thing is to ask what the problem is. Has an analytical method you've been using suddenly stopped working? Do you need a theoretical equation to analyze the data you've accumulated? Have you run across an observation that, although repeatable, simply doesn't fit with the prevailing understanding of how this system should behave? You may not be transforming the domain, but you're faced with the creative challenge of understanding of your research tools and changing your own research environment.

In this chapter we'll deal first with creativity, thinking of it as the contribution you'll make to scientific knowledge as a result of your doctoral or postdoctoral research. Then we'll discuss problem solving and troubleshooting, the day-by-day puzzles you'll have to work through to make progress on your research. And it should become clear that creativity and problem solving require similar ways of thinking.

The creative process

Studies describing the creative process have usually broken it down into steps, something like this:

- Preparation. Becoming immersed and thoroughly familiar with a problem, bringing prior knowledge to bear, identifying what appear to be the key questions and what would constitute satisfactory solutions.
- Incubation. Stepping back from the problem, letting subconscious processes take over.
- Insight. Either the full answer or a critical element, comes to the surface of consciousness. This is when one exclaims, "Eureka!"
- Verification. Checking whether the illumination really is a satisfactory solution to the problem.
- Elaboration. Working out the details, implications, and applications of the solution. This can take the most time, since in some sense you are embarking on a new endeavor.

The creative process is almost never as linear as described here. With scientists of great insight and experience, the first three steps may merge into an apparently seamless intuitive process. And there are always further periods of new insights, circling back, rephrasing the basic question, noting more implications, finding new ways to verify, involving both conscious and unconscious mental activity.

Asking the right questions

The first step of creativity in research entails seeing that there is a problem, noting that it is indeed something worth investigating, and formulating a question. Sometimes there's a real puzzle: Why does the universe behave as if it has ten times as much mass as we can detect? How can such a complicated system as the eye possibly have evolved? How do we remember things?

Sometimes the most creative insights come from asking questions that other people haven't even thought about. A famous example is Einstein's wondering what an observer would see if he could ride a beam of light, which led to the special theory of relativity.

At other times there's a dearth of information and a recognition that filling in the gaps would probably be useful: How many species of plant are there in this patch of jungle? How many children live in poverty in Minnesota? The creativity of cataloging is often doubted. In biomedical science, for example, there has been a prejudice in favor of "hypothesis-driven research." The Human Genome Project, by that standard, was simply a mechanical accumulation of gene sequences, with no questions being asked and no hypotheses being tested. On the other hand, as the human and other genomes have been sequenced, all sorts of surprises and questions have emerged based on the data that have been uncovered. Why do human beings have fewer than twice as many genes as the flatworm *C. elegans*? Why are humans and corn 80 percent genetically identical? In a different domain, statistical accumulation of causes of death led to the realization that the majority of deaths in the United States are due to behaviors (overeating, drunken driving, smoking, homicide) rather than diseases, which has led to very difficult questions about how to change behaviors—a task much more difficult than vaccinating every child for polio.

And at still other times, when trying to troubleshoot a piece of equipment or figuring out the proper statistical analysis of a data set, the question poses itself: What's gone wrong with this equipment? Which test, out of the many described in this statistics textbook, is appropriate for my situation?

Creativity requires breadth

CONCEPTS FROM DIFFERENT AREAS

Arthur Koestler, in his book *The Act of Creation*, defines creativity as the bringing together of concepts from two different areas. This may not be a universal mechanism of creativity, but it contains an important germ of truth. If the standard methods and concepts of a field are inadequate to explain some important phenomena in the field, then perhaps an idea from outside will do the trick. It doesn't have to be something new in the other field. Just the idea of bringing it into the first field, and making it fit, is creative. Consider, for example, bringing together the concept of gels that swell strongly when they absorb water with the concept of diapers. This led to the invention of diapers that absorb water not just in their fabric but in a small amount of gel engineered into the structure of the diaper.

To make progress on your project, beyond straightforward data-gathering, you will have to introduce something new to what is already being done in the field. This may be a new technique, a new way of preparing your materials, or a new theory to explain the results. "New" in your field probably doesn't mean totally new to science. It's very likely that an appropriate analytical method, theory, or concept has already been developed in a neighboring field. All you need to do is to become aware of what's available and apply it to your problem.

BREADTH OF KNOWLEDGE CAN COME IN SEVERAL WAYS

Talk with your professor. Ask about the significance of her work and its implications for neighboring fields and the rest of science. This may be one of her first services as a mentor, giving you advice and perspective about a fairly large piece of scientific territory. In exchange for your apprenticeship, she can give you a sense of the broad world of science, the beginnings of the feeling of being in a network of intellectual relationships.

Talk with people outside your research group. Find out what they're doing, and think about how it might apply to your research problem. Explain your work to them; maybe they know something, or someone, that could help.

Go to seminars outside your field. A physical biochemist can learn about interesting problems, and new ways to design and analyze biological molecules, from a molecular biologist. Conversely, a molecular biologist can learn about new ways to conceptualize problems, and techniques for more detailed molecular analysis, from a physical biochemist.

Browse the literature. Spend some time reading outside your field, particularly review articles written for a broader scientific audience: *Scientific American*, *Trends in Biochemical Sciences*, *Essays in Biochemistry*, and reviews in *Science*, *Nature*, and *Physics Today*, among others. If you pick up a scientific journal to read a particular article, take a few minutes to browse the rest of the issue.

The danger in seeking breadth is becoming a dilettante, never focusing on the most important project. To avoid this, seek breadth actively and purposively. When you read or go to a lecture outside your field, ask yourself, "How can this be applied to my project?" Try to spend your "breadth time" in ways that seem likely to be fruitful.

It's good to be open to unexpected new ideas There are many famous instances of serendipity—occasions in which scientists have come across crucial new ideas when least expecting them. But remember, as Louis Pasteur said, "In the fields of observation chance favors only those minds which are prepared."

SCIENCE CHANGES RAPIDLY

Another reason for broadening your knowledge and interests is that science advances quickly. The tools of science are so powerful, and there are so many bright people using them, that the hot problems of the moment may seem quite routine in a few years. Because of shifts in social and political priorities, availability of grant funding to pursue certain types of research may also change dramatically.

Five years from now, the research question that seemed so interesting when you first asked it may have been answered (hopefully by you) or may have reached a dead end. What do you do then? If your interests have stayed broad, you will be aware of other interesting problems and be prepared to tackle one of them. In fact, after five years of plugging away at a narrow project, you may be thoroughly tired of it and eager to try something new.

Optimizing the conditions for creativity

Students of creativity have noted that certain conditions encourage innovative problem definition and creative problem solving. In this section we list some of the most important.

AIM TO BE CREATIVE EVERY DAY

Creativity is not just a matter of random inspiration; it involves steady work and thought. In your work, you face both large and small problems. Approach both creatively. Start each day with a problem to solve. In the words of John Platt, try to do a "gamesworth" of genuine scientific thought each day—the same amount of thought you would put into a game of chess or a hard crossword puzzle. This is a different kind of activity than routine experimental or computational manipulations.

FIND CREATIVE COLLEAGUES

Perhaps the single most important support for creativity is having creative colleagues. There is a critical mass effect in being surrounded by bright, creative people. Mihaly Czikszentmihalyi says, "In the sciences, being at the right university—the one where the most state-of-the-art research is being done in the best equipped labs by the most visible scientists—is extremely important. George Stigler describes this as a snowballing process, where an outstanding scientist gets funded to do exciting research, attracts other faculty, then the best students—until a critical mass is formed that has an irresistible appeal to any young person entering the field" (1991, 54–55).

Outstanding colleagues are both supportive and critical. They will encourage you to do your best, ask the tough questions that will lead you to think more deeply about your research, and set a good example of high achievement.

Even without a large group working in the field, a key collaborator can be very important: think of Watson and Crick. Having at least one other person to bounce ideas off can do wonders for your creativity.

ARRANGE CREATIVE SURROUNDINGS

To the extent that you can, arrange the space in which you do research so it's comfortable for you, so that you feel in control of your environment. This may be difficult if you have a desk at the end of a lab bench in the middle of a big lab or if you share a small office with several other people, but do the best you can. If you like a space that's neat, with everything filed away except what you're working on at the moment, then make sure you have file cabinets and bookshelves. If you like to work in the midst of "creative disorder," feel free to do that, as long as you can lay your hands promptly on what you need. If you like pictures and mementos on the walls and shelves, feel free; if you like a Spartan atmosphere, that's OK too. If you like to work to music, do so—but if you share the space with others who might have different tastes, use earphones. If you like quiet, and others are disturbing you with their music or chatter, ask them to pipe down or use earphones for their own listening enjoyment. Try to control the climate so that it's neither so warm that you feel sleepy nor so cold that you feel uncomfortable all the time. In other words, try to arrange the space so that it serves your purposes and you can come to think of it as the place where you get good, enjoyable work done.

TAKE BREAKS

You can't do experiments or computer programming or reading or math calculations all the time. Eventually you'll get stale and tire of continual concentration. Your productivity will go way down, and the amount you'll accomplish in your long days will diminish. Give yourself a break every once in a while. Physical activity is particularly good. Several notable writers (Wordsworth, Tennyson) and scientists (Fermi, Heisenberg) were famous for taking long walks. Swimming, biking, tennis, softball, Frisbee, working out at the gym, and other physical activities take your conscious mind off your work, while you incubate creative ideas subconsciously. Relaxing the mind after a stint of hard thinking is one of the best ways to generate the creative ideas that have been eluding you. Physical activity also is good for your health and, if you do it with others, for your friendships and social life.

While physical activity is important, other pastimes—playing music, taking photographs, painting, cooking—that divert you without using too much of the verbal and rational parts of your brain are also good for letting creative ideas incubate and simmer. Do not, however, give in too much to thoughtless temptations such as watching TV or drinking. If you find yourself tired and stressed at the end of a hard day, the best thing for it is a good night's sleep.

MAINTAIN VARIETY

There are many activities associated with research, and one way you can keep yourself from getting stale is to move among them. Here, for example, is Czikszentmihalyi's description of the work habits of the famous ecologist and sociobiologist E. O. Wilson:

> Wilson typically works on several projects at once, using different methods. This is again a common pattern among creative individuals; it keeps them from getting bored or stymied, and it produces unexpected cross-fertilization of ideas. There are at least four different approaches that Wilson uses. The first is fieldwork in exotic places, which acts as a sort of "nuclear fuel" by providing concrete experiences and data to be elaborated later. The second is attending lectures or meetings, where he absorbs from other experts the latest developments in the domains that interest him. The third is night-work, the serendipitous connection between ideas that unexpectedly arise upon waking up in the middle of the night. And finally there is the systematic work that takes place from morning to early afternoon, which also includes reading, writing, mathematical modeling, and drawing specimens. The crucial insights sometimes occur during the night-work, but more usually they are the result of the systematic work process itself and its combination with the other three approaches. (1991, 272)

Long ago, researchers were much less specialized. As a scientist, Ben Franklin, for example, was mainly interested in electricity. However, he dabbled in many areas and made significant contributions to all. He invented swimming fins, bifocals, a glass harmonica, watertight bulkheads for ships, the lightning rod, an odometer, and the wood stove (called the Franklin stove). One has to wonder if his wide-ranging research interests fueled his creativity by linking key pieces of knowledge scattered in different scientific disciplines.

Even if you're a beginning grad student and have only one project in the works at once—and little opportunity to travel to exotic places—you can still move from bench research to reading, writing, statistical analysis, plotting graphs, or going to lectures, and then back to the bench.

Creative use of tools

FINDING AND APPLYING THE BEST TOOLS

Science makes progress at least as much through the use of new tools as through new concepts. Tools can be experimental, theoretical, computational, or informational. Just as in any endeavor in which tools are used, the right one can save time and effort, and can lead to major improvements in quality and precision of the result.

In your research you will want to apply the most powerful appropriate tools to your problems. Be adventurous in trying new experimental techniques. Don't just be satisfied with the techniques currently in use in your lab. In biochemistry and molecular biology, for example, a variety of experimental tools—instrumental, chemical, biochemical, and genetic—are all important. Similarly diverse sets of approaches are available in other scientific disciplines. Think about ways each of them might be applied to your project. Learn what tools other researchers, and not just those in your department, are currently using to solve similar problems. A large research university has scientists in various departments using many of the most modern techniques. They are generally willing to help, and perhaps to collaborate. A telephone call, or a short trip down the hall or across campus, can get you started. It will be worth your while to devote time systematically to learning about new techniques and thinking about their relevance to your problem.

Creativity in tool development is one of the most characteristic aspects of scientific research. Sometimes an existing tool is not sensitive or specific enough, or doesn't have enough resolving power. Sometimes it's not efficient enough for large-scale data collection, so new types of sample processing and automated signal detection need to be developed. The development of high-throughput DNA sequencing methods for the Human Genome Project is a dramatic example. Sometimes an entirely new type of detection is developed, either purposefully or accidentally. Donald Glaser, for example, used his casual observation of the formation of bubbles in a glass of beer as the basis of the bubble chamber used to detect of subatomic particles, an invention that revolutionized particle physics in the 1950s and led to his 1960 Nobel Prize.

Sometimes the right tool is already in use in another field and needs only to be brought over. Jack Oliver, in *The Incomplete Guide to the Art of Discovery*, writes, "This guideline describes what is probably the most consistently successful way to make new discoveries in an observation-oriented branch of science. . . . The trick is simply to bring instruments and measurement techniques from one branch of science into a different branch for the first time. . . . This approach . . . is so effective because new kinds of observations of an important scientific phenomenon nearly always produce surprise and discovery" (1991, 66–67).

THE ROLE OF THEORY

Theory is also a tool. It can help you plan experiments, predict and understand results, test ideas, and generate new experiments. A prime virtue of theory is that it gets you in the habit of thinking quantitatively, and thus thinking more analytically about your experiments. Completely rigorous theory is generally impossible in complex fields such as biology, but even approximate theories and crude calculations can be very useful. Before you start an experiment, you should make a quantitative estimate, even if only a crude, "back of the envelope" one, of what magnitude of effect you expect if your hypothesis is true. This can tell you whether the effect is likely to be observable, and will help you focus better on the meaning of your result once you have obtained it.

Never be satisfied with just having measured or observed a phenomenon. Try to connect it to a theoretical framework. The most satisfactory theory is quantitative, but this is not always possible in the initial phases of an investigation. A qualitative conceptual or mechanistic scheme can be equally important in planning and interpreting experiments. But it is usually possible eventually to make such schemes quantitative, which enhances their predictive and explanatory power.

STATISTICS

Statistics is a major tool in many branches of science, including agricultural, medical, and survey research, and could probably be used profitably even more frequently. As scientists work with smaller and smaller samples—increasingly even with single molecules—or try to extract small but significant effects from a large and noisy background, statistical analysis becomes essential. Many physical and biological scientists have insufficient training in statistics, so you might gain an advantage by learning and using statistical tools in your research.

While statistical analysis can reveal small yet seemingly significant effects, it's important to try to connect these effects to the underlying reality. A tiny change in cell function, for example, might be statistically significant but have no biological significance. "There are three kinds of lies: lies, damned lies, and statistics," said the British statesman Benjamin Disraeli (or possibly Mark Twain—the attribution is uncertain). Perhaps more to our point, Scottish author Andrew Lang once wrote, "He uses statistics as a drunken man uses lamp-posts—for support rather than illumination."

Techniques for creative problem solving

In this section we list a number of tried and true approaches to creative problem solving. They are complementary rather than mutually exclusive: a tool kit for creativity.

KNOW THE FIELD

As we've already said, formulating a productive question is the first step in creative problem solving. The next step is to bring your knowledge base to bear. You can't answer difficult, field-specific questions in a vacuum or by pure reason. You need to know what is known in the field that's relevant to the question. What pertinent information is missing? Are there some things that are accepted as true by most workers in the field that in fact rest on shaky foundations? Your graduate-level courses, seminars, and independent reading, as well as conversations with experts, will give you this knowledge of the domain. It's a never-ending process, continuing through life.

BRAINSTORM

If a systematic approach doesn't work, you may want to try brainstorming, sometimes called "divergent thinking." The key here is to produce as many different ideas as possible, not censoring yourself by immediately rejecting ideas that seem unlikely or even outrageous. Of course, there must eventually follow a period of sifting and winnowing, but try for a time to put no restrictions on your idea generation. Numerous great scientists have said, in one way or another, that the secret of their success was generating ten ideas and only keeping one.

DRAW DIAGRAMS

Think visually as well as verbally and mathematically. Draw diagrams or mind-maps that illuminate connections, illustrate cause and effect, show feedback or feed-forward loops, or display hierarchies.

ARGUE BY ANALOGY

Sometimes a hypothesis or way of thinking about a problem can be generated by drawing an analogy to a more familiar situation. You may see similarities of pattern or approach, even though scales of time and space may be very different. Let your imagination run loose—that is, again, brainstorm—to find useful analogies.

FORMULATE A FERTILE HYPOTHESIS

A hypothesis, no matter how unsupported or seemingly arbitrary, can be an invaluable aid to structuring and focusing your thinking and exploration. It can help you coalesce and relate a wide range of seemingly unrelated results and musings, and can lead—as you see how they support or contradict the hypothesis—to a productive evolution in your thinking. Here are relevant quotations by two people who have thought deeply about creativity:

> The shrewd guess, the fertile hypothesis, the courageous leap to a tentative conclusion—these are the most valuable coin of the thinker at work. —Jerome S. Bruner (1960, 14)

> Scientists who are timid about proposing a fresh new hypothesis because the evidence has not forced them to do so are not likely to achieve the big discovery. —Jack E. Oliver (60)

DEVISE MORE THAN ONE TESTABLE HYPOTHESIS

Although a fertile hypothesis can be both productive and energizing, it can also be misleading. It's best to avoid emotional overinvestment in any single idea, hypothesis, or solution. If possible, generate at least two alternative hypotheses, then convert them to testable form and proceed to test them. Don't do experiments just to see what happens. Do experiments to distinguish between your hypotheses or to demonstrate the falsity of one of them. In the usual description of the scientific method, one generates a hypothesis, tests it against observations or experiments, revises the hypothesis, and tests again. Our approach is slightly different. We encourage you to start with several competing hypotheses and to recognize that there is a psychological, intuitive, even illogical and emotional, aspect to the cycle of hypothesizing and testing.

How you define and view the problem will have a strong influence on what you think a solution may be. Try to examine it from many different points of view. Each viewpoint will suggest a different formulation and different ways of testing your ideas. It may take more time to generate and examine several alternatives, but the end result will be better for it.

DON'T REACH CLOSURE TOO SOON

A dangerous temptation in dealing with a difficult problem is to cling to the first solution that comes to mind. Be careful! Most difficult problems have subtleties, and a wide variety of conditions that must be satisfied simultaneously. A good correlation between experimental parameters does not necessarily translate into a cause-and-effect relationship. Being wedded to a premature conclusion might blind you to other possibilities. The first solution you come up with may continue to hold water all the way through, but it may not. Have a backup plan if the idea doesn't prove correct when tested further.

VARY YOUR THINKING STYLE

Don't get locked into a single way of thinking about a problem. Different thinking styles are appropriate at different stages. At different moments, you may want to

- Make back-of-the-envelope numerical estimates of important quantities and effects, especially if you don't have all the data you'd like.
- Listen to your hunches, intuitions, and feelings.
- Apply judgment, list reservations, and think about why a particular approach won't work.

- Be optimistic about why things will work and what benefits will ensue.
- Generate lots of ideas and many possible, and hopefully provocative, alternatives.
- Alternate between thinking about the big picture and breaking the problem into small steps, each of which might be more readily solved.

In the back of your mind, however, you should maintain an overview, remaining conscious of which approaches you are bringing into play and assessing whether you are doing so at the appropriate times.

Polya's heuristic

The mathematician George Polya, in his well-known book *How to Solve It*, summarizes these ideas about creative problem solving under the heading of "heuristic": "The aim of heuristic is to study the methods and rules of discovery and invention. . . . Heuristic reasoning is . . . not regarded as final and strict but as provisional and plausible only. . . . [B]efore obtaining certainty we must often be satisfied with a more or less plausible guess" (1971, 112–13). Polya outlines the steps with reference to solving mathematical problems, but the principle carries over with little modification to other realms of creating problem solving.

- Be sure you understand the problem: the unknowns, the conditions, the data given.
- Devise a plan by thinking of a similar or related problem, perhaps simpler or more specialized or more general, that you know how to solve.
- Carry out the plan, checking your steps along the way.
- Review your solution, seeing whether you could have solved the problem in a different way, and thinking whether you could use the result or approach in a different context.

Troubleshooting

In the hierarchy of scientific discovery, the creative leap that illuminates or changes a whole field comes at the top, followed by the posing and working out of research problems that incrementally move the field forward within its previously established boundaries. In day-to-day work, however, there is a third level: troubleshooting the things that should work but don't—the instrument that's giving the wrong reading, the reagent that's not reacting, the computer code that's crashing. Here are some suggestions on how to deal with these annoyances.

ASK OTHERS

What should you do when you run into a problem you can't resolve? The wry old rhyme, "When in panic and in doubt, run in circles, scream and shout," isn't a useful response. But letting a select portion of the world know about your problem may be the right step. If you have a question about what to do next, and the obvious things haven't panned out, then asking for help from somebody who might know the answer or have a good idea is a reasonable strategy. Ask your labmates or advisor for ideas, or bring the question up at group meeting. There might be a lab down the hall, or across campus, that has expertise in a relevant area. Or you might try to contact an expert at another university by e-mail, phone, or letter.

All of these approaches are OK, though making your initial inquiries close to home is probably best—to avoid imposing unnecessarily on the kindness of strangers or prematurely giving a competitor an idea of what your lab is working on. It's obviously easiest; people in your own lab will know what you're talking about without too much explanation, whereas an outsider would require a more thorough and explicit description of the problem and what you've already done to try to solve it. That process of formal description, however, may well turn out to be a useful part of the process, one that reveals the elusive solution by making obvious something you've overlooked or leading to an aha! moment. Even if you don't turn to an outside expert, you should certainly record such a description in your notebook.

CALL THE MANUFACTURER

In these modern times when you buy instruments and reagents and software from companies rather than preparing them yourself, most companies offer free technical help. Use it. If no colleague or scientist you talk can come up with a useful answer, then call the manufacturer. Instrument companies typically employ technical experts who are quite willing to suggest key points to check or calibrations to run; they can also tell you which parts are most likely to cause trouble. If you're not expert in the electronics or optics that lie within today's wonderfully sophisticated black boxes, then you might want to enlist your local electronics shop technician as a translator. (You might also approach the shop for help first.) Likewise, if you run into problems with the preparation of an enzyme (notoriously fragile and finicky beasts), calling the supplier will probably put you in touch with a staff biochemist who is familiar with details of the molecule's behavior and can clarify what its response to pH, temperature, or solvent conditions might be.

BORROW FROM OTHERS

If the problem isn't a malfunction of something that was previously working or should have worked, but rather requires a new approach, then what do you do? If your assay simply isn't sensitive enough, or quick enough, or specific or discriminating enough, then you'll have to find something better if you're to be able to pursue your project in the way you had hoped. Again, the first step is probably to ask around. Maybe one of your contacts knows of a better tool. Or you might search the literature for related assay methods. (See chapter 16 for tips on searching the literature.) Say you find something that looks interesting and potentially suitable in the literature, but it's too expensive for your lab to buy, or it would take too much time to get. Perhaps you can arrange to visit a lab that already has one, or get a demo from the manufacturer, and at least see whether it will serve your purpose. If it does, perhaps you can arrange for longer-term use, which may involve spending time at a distant lab. This in turn may require a formal collaborative agreement, with the director of the host lab and perhaps the student or postdoc or technician who helps out receiving joint authorship on your paper. Or perhaps an acknowledgment will suffice; work this out ahead of time to stave off ill feelings later.

TROUBLESHOOT SYSTEMATICALLY

If you're doing experiments with complex apparatus, understanding its workings and being able to troubleshoot it are important. It's a corollary of Murphy's Law that the equipment is most likely to start showing hard-to-diagnose symptoms just as you're ready to start a key experiment. Some pointers (after chap. 6 in E. B. Wilson, *An Introduction to Scientific Research*):

- Don't ignore the beginnings of strange new behaviors, even if they seem minor. They may betoken the onset of serious trouble. Try to understand what's going on. Have faith in the law of cause and effect, and the virtues of a systematic approach. Don't forget the possible effects of aging: corrosion, batteries needing replacement, etc.

- If you're modifying the instrument or changing operating procedure, change just one thing at a time. That way, if things start to go wrong, you can get back to solid ground. If things are working well, don't change them, particularly just before a crucial experiment.
- Make sure things are working well before a crucial experiment, by going through all the steps with well-designed controls. Whenever possible, include positive and negative controls. Be sure the instrument is properly calibrated.
- If there are multistep procedures to be followed, make a checklist and follow it.
- To the extent possible, analyze your data as you go along. This will enable you to spot unexpected behavior and either correct the experiment or follow up on a potentially important new finding while everything is up and running.

Facing success and failure

Most of the time, the approaches we've discussed in this chapter will take you over the obstacles, large and small, that confront you as you pursue a research project. Sometimes, however, a problem will just not yield to normal approaches, patience, and hard work. On rare occasions that will be because you've discovered something quite unexpected, in which case congratulations are in order: your career is off to an exciting start. Other times—more common, unfortunately—the project was not well conceived and cannot be completed as originally defined. What then? Here are a few ideas.

DEALING WITH REAL SURPRISES

What happens if you obtain really unexpected results that seem to be reliable and reproducible but are not explicable in terms of the current understanding of the field? This is where "real" creativity comes in, of the sort that changes fields and makes reputations. There are at least two approaches, different but not mutually exclusive. The first is to generate hunches and devise experiments to prove or disprove them. The second is to go back to first principles, examine and question as many assumptions as possible, and ask which of them may not apply in this instance since, although most people accept it, it is not supported by much evidence. You will want to generate lots of ideas, making lists and drawing diagrams freely, without much internal editing, to dredge up unconscious thoughts that your internal editor would screen out if you gave it a chance. Lots of incubation—thinking hard about the problem but then letting go and giving time a chance to do its work—is probably necessary here.

REDEFINE THE PROJECT

If extensive work on a problem, using all the tools we've listed, does not help, you may need to look in a different direction. Sometimes the initial conception of a research project just doesn't work, for reasons that weren't apparent when you started. At some point, you may need to cut your losses and move on to something that's more feasible. Assuming you stay in the same lab, the new project will probably be related to the old one, but you may need to change to a different organism, do a survey rather than a detailed mechanistic study, or use a less sensitive but more stable analytical technique. Such is the uncertainty of scientific research. However, the thought and effort that you've invested are rarely wasted. You've explored and learned a lot, and what you've learned in the failed first project will probably make the second one more successful.

Take-home messages

- Train yourself to appreciate the various steps involved in the creative process: preparation, incubation, insight, verification, and elaboration.
- Surround yourself with creative colleagues.
- Ask research questions that are both novel and answerable.
- Whenever possible, frame your questions in the form of a testable hypothesis supported by strong background information.
- Aim for research questions that bridge different areas. Exciting new knowledge often lies at the boundary between disciplines.
- Keep your knowledge up to date.
- Keep at the cutting edge of research technology.
- Try to design multiple approaches to answering each research question.

References and resources

Booth, Wayne C., Gregory G. Colomb, and Joseph M. Williams. 2003. *The Craft of Research.* 2nd ed. Chicago: University of Chicago Press.

Bruner, Jerome S. 1960. *The Process of Education.* Cambridge: Harvard University Press.

Csikszentmihalyi, Mihaly. 1996. *Creativity: Flow and the Psychology of Discovery and Invention*. New York: Harper Collins.

Koestler, Arthur. 1990. *The Act of Creation*. Penguin.

Oliver, J. E. 1991. *The Incomplete Guide to the Art of Discovery*. New York: Columbia University Press.

Polanyi, Michael. 1974. *Personal Knowledge: Towards a Post-Critical Philosophy*. Chicago: University of Chicago Press.

Polya, George. 1971. *How to Solve It: A New Aspect of Mathematical Method*. 2nd ed. Princeton, NJ: Princeton University Press.

Wilson, E. Bright, Jr. 1990. *An Introduction to Scientific Research*. New York: Dover.

14 STAYING MOTIVATED

Motivation is key to success in scientific research, as in most other endeavors. Research is hard work, with many disappointments. Motivation keeps you going and makes the work fun. Motivation generates energy. If you are really enthusiastic about your work, you remove the internal conflicts that waste energy. This chapter describes ways to maintain or regain motivation.

Motivation and flow

The years of graduate school and postdoctoral studies can be some of the best, most rewarding times of your life. This is your opportunity to become completely absorbed in a field of study and research, a situation that the psychologist Mihaly Csikszentmihalyi has called "flow":

> Flow—the state in which people are so involved in an activity that nothing else seems to matter; the experience itself is so enjoyable that people will do it even at great cost, for the sheer sake of doing it. (1991, 4)

Csikszentmihalyi observes, "The best moments [in our lives] usually occur when a person's body or mind is stretched to its limits in a voluntary effort to accomplish something difficult and worthwhile" (1991, 3–4). Such efforts are often painful at the time, because they involve an all-consuming struggle to overcome challenges that we have consciously decided are very important to us. To embark on such a struggle is neither easy nor natural; it requires strong, persistent efforts of will. But eventually the feedback between effort and enhanced skill and understanding takes over, and the resultant sense of inner reward makes the process enjoyable. If you're lucky and have chosen well, this sense of intrinsic accomplishment and inner reward may persist throughout your professional life, even though you may have chosen the profession for extrinsic reasons—to help others or to assure a good living.

Csikszentmihalyi lists nine characteristics of the flow experience:

- There are clear goals every step of the way.
- There is immediate feedback.
- There is a balance between challenges and skills.

- Action and awareness are merged.
- Distractions are excluded from consciousness.
- There is no worry of failure.
- Self-consciousness disappears.
- The sense of time becomes distorted.
- The activity becomes autotelic (something which is an end in itself).
  (1996, 111–13)

Competitive sports are probably the most prominent examples in our society of flow-inducing experiences, but those who have experienced it recognize that high-quality scientific and scholarly research can be equally absorbing. The daily work of science, as well as its more dramatically creative moments, can be made occasions for flow: be creative in routine analytical work, aim for innovations in precision and economy, strive for elegance in routine manipulations.

Flow can be both an individual and a group experience: the interactions of a highly achieving, productive, cooperative but competitive lab or department are not so different from those of a championship sports team. The culture and society of science provide fine opportunities for flow.

Motivational difficulties and distractions

Despite the great psychological rewards research offers, staying motivated is a problem for graduate students and postdocs. It may be that your project is taking longer than you think it should or just isn't being successful. You've been working too hard on it, not getting enough sleep or exercise or recreation, and you're burning out. You may have too much to do, or too little information, and be unsure what to do next. Those around you, students and faculty alike, may seem impossibly accomplished—way out of your league. Or, as you try to write your dissertation or a journal article, you may run into writer's block. Altogether, your confidence and self-esteem are suffering, and you're questioning whether this is really something you want to put your time and effort into.

In other cases, distractions may be preventing you from maintaining the motivation and deep involvement that you need to pursue your research successfully. Distractions come in many forms—financial worries that push you to take an outside job, difficulties in personal relationships, experiencing sexual harassment or discrimination—and can reduce motivation in at least three ways:

- First, distractions are often attractive, so they compete with your real interests—or they may be other real interests, competing on a conscious or subconscious level for your attention.
- Second, distractions, if attended to, take up time and thereby slow your progress. Thus your goal begins to seem farther away, less attainable, and loses some of its appeal.
- Third, distractions use up energy, leaving you less able to do the important things when you're most pumped up to do them.

Keys to high motivation

Having noted the importance of motivation, and some of the things that can undermine it, let's take a look at some ways by which motivation can be built and maintained.

BUILD AND MAINTAIN SELF-CONFIDENCE

First of all, keep a positive, optimistic attitude. You have every reason to believe that your project is doable and worth doing, and that people no smarter than you have solved similar problems. A steadfast, systematic, upbeat approach will bring you to a successful conclusion.

SET GOALS AND PRIORITIES

Flow happens when you're working on challenging, absorbing problems of your own choosing. You'll be energized when you have a clear view of where you hope to go and what the most important steps are to get there. Reaching the point where you can publish a few good papers in good journals is the sort of goal you will presumably set. Don't ignore other responsibilities—teaching, personal life, and so on—but keep your eye on your main goal as you set each day's priorities.

BREAK DOWN THE TASK

Sometimes you may have trouble getting down to work because, subconsciously, you don't quite know what you should do next. At such times, more explicit, detailed planning may be in order. Make a list of all the things you can think of to do: Searching the literature? Consulting with an expert? Locating new equipment? Calibrating, or making arrangements to use, existing equipment? The more concrete, finite steps you can identify, the easier it will be for you to get started on one of them. This is sometimes known as the "salami principle": slicing up a big task into a series of smaller, less intimidating steps.

IDENTIFY THE NEXT ACTION

Personal productivity guru David Allen, in his book and Web site called *Getting Things Done*, points out the importance of identifying the next thing you can actually do to move your project forward. If you'll be presenting a poster at a national meeting in six months, and you're paralyzed thinking about how much you need to do to get ready, ask yourself, "What's the next action I need to take to move this along?" You might consult the meeting announcement or Web site to find out the deadline for submitting the abstract for the poster, and to find the physical size of the space into which your poster will have to fit. Or you speak with a labmate who's working on a related project about how to divide up the presentations. Each of these is a small task, but it takes you a step closer to your goal, gives you a feeling you're making progress, and gets you more involved with the project. Sometimes a little push is all that's needed to start the engine. Make some notes on the day's work. Arrange your materials. Turn on the computer and call up the file you have to edit. Do something physical to get you moving in the right direction.

GET MORE INFORMATION

Sometimes the appropriate next action is to get more information. Your lack of motivation may arise from not knowing enough about the background of your project (this may particularly be true at the beginning). Alan Lakein writes, "Not until a certain level of familiarity with a subject is reached are people likely to push toward a further exploration to satisfy the newly aroused curiosity. Once that level is reached, there is a good chance that involvement will increase as knowledge accumulates—which is why getting more information is an instant task that leads to involvement" (1973, 111). Or you may be missing a key piece of information that is needed to make a specific plan; this too can lead to inaction. In either case, read, search online, or talk to someone until you get the information you need.

On the other hand, sometimes you have so much undigested information that you're overwhelmed and paralyzed. Now it's time to make sense of what you've got. Take each item in turn, and ask yourself, for example, Is this new, or do I already have it? Is it useful? How might I classify and file it? Can I summarize it, rather than keeping the whole thing? Write things down as you go, so that you have a permanent record of your thoughts. See chapter 16, on managing information, for more details.

GET SMALL RESULTS EARLY AND OFTEN

It is psychologically important to get results early in your research career, even if they are not original. Even small, unoriginal results are valuable in developing a habit of accomplishment. Rome wasn't built in a day, nor can your research be completed that quickly. But recording a small but solid result each day—say, mastering a technique and being able to reproduce a result in the literature—can give you a sense of confidence. Here, as we noted in chapter 10, your notebook can serve as a source of motivation. As John Platt writes, "The notebook satisfies the mind as a record of achievement, as a pleasure to reread and criticize, and so at last a pleasure to anticipate adding to" (1962, 139).

ESTABLISH A ROUTINE

Motivation is easier to maintain if you can spend much of your time on autopilot, not worrying about to do next. A routine can eliminate many internal psychological conflicts. This can be as simple as getting to the lab by 8 each morning, spending the first hour planning your day's work in your notebook, and writing up the day's results and ideas for tomorrow in the last hour before you leave at 5:30, knowing that you'll go to a departmental seminar on Wednesday afternoon and the grad student journal club Friday at lunch.

Your routine should include major blocks of time during which you will not be distracted. Distractions are unavoidable to some extent. There are phone calls and e-mail to attend to, people to talk to, family matters that demand attention, routine duties such as meetings, filling out requisitions, and preparing reports. Nevertheless, it's important that you set aside a sizable block of time, free from distractions, to work on your main thing. Shut the door, and don't open e-mail or browse the Web, until you've done your allotted chunk of work. Then you can attend to routine chores or a bit of recreation before moving on to the next piece of real work.

VARY YOUR ACTIVITIES

Routines are great, but they can become stale and unproductive after a while. It's hard to keep going on the same repetitive tasks, hour after hour and day after day. Fortunately, research involves many different activities, and when you find one getting stale you can switch to another. As Alan Lakein writes:

> It's not at all surprising that you should become bored, restless, or fatigued after working on the same task for some time. The yearning for change is natural. . . . Normally your need for stimulus change is satisfied by the evolution of the current task as one step leads to another. The natural evolution of a project generates many possible tasks. Keeping a To Do List gives you a choice of tasks to provide stimulus change. It even pays occasionally to take time to list additional steps you might take on just so you have a broader range of activities to draw upon. . . . What works best is to change the way you're going about the task. Any intellectual effort contains elements that can be juggled. You need information and ideas. These have to be collected, digested, and acted upon. You can shift back and forth between working on information and ideas, or switch from collecting to digesting to action. (1973, 120–22)

Sometimes it helps to vary your location. If you're trying to write at your desk in the lab and the words aren't flowing, take your laptop or notebook to the library or the local coffee shop.

You will also find that your moods shift with time, and when you don't feel like doing one activity another may seem more appealing. If you feel chatty, wander down the hall and talk with a friend about your project. If you're annoyed because your desk is piled with papers, spend a little time sorting and filing them. Or if you're feeling momentarily grumpy and discouraged, give yourself a pep talk or make a commitment to someone that, by gosh, you're going to finish this little piece of work by the end of tomorrow.

TAKE A BREAK

Try not to sit at your desk, staring at your computer screen for hours at a time. A mind locked into a mindset is not going to be in the best shape to think new thoughts. Give your mind the space to get unlocked. Wandering down the hall to get a drink of water or browsing in the library may be all that's needed. Find some way to remove yourself periodically from close combat with the unyielding problem.

Even when the project is going well, you may feel burned out and unmotivated, so give yourself some free time. Consider it a reward. Go home early, go out with friends, get some exercise, go to a play or musical event, get a couple of good nights' sleep in a row. Perhaps this would be a good time to take a long weekend, go skiing, or otherwise get some distance from the project. Mental muscles, just like physical ones, get tired and need time to discharge their metabolic wastes and generate new stores of energy.

GET ENOUGH EXERCISE AND ENOUGH SLEEP

The Council of Graduate Students at the University of Minnesota offers some wise advice in its booklet "Staying on Course: Mutual Roles and Responsibilities in the Graduate School Experience": "Keeping yourself healthy (physically and emotionally) should be your number-one priority. If that means 30 minutes at the gym every day plus pottery twice a week, so be it. And it goes without saying, get enough sleep whenever possible (every night!). Late nights in the lab or poring over journals don't leave you in any condition to think constructively about your progress and your future."

BUILD AND USE YOUR SUPPORT GROUP

Doing science can be lonely. Creativity is full of self-absorption and false starts. Your advisor should bolster your self-confidence and enthusiasm by helping you choose a problem that's significant but doable, by setting things up so that an initial string of successes is likely, by supporting a switch to a different problem if the first proves undoable or poorly suited to your talents, and by fostering an atmosphere of interest, enthusiasm, and connection to the broader scientific world within his or her research group. (An advisor's reputation for enthusiasm and support is one of the most important things to check out when you're deciding which research group to join.)

But don't rely just on your research advisor. Have other mentors (faculty members, senior grad students, or postdocs) to whom you can talk about research progress, personal conflicts, and career issues that may be affecting your motivation. You should also try to find people outside your research area with whom you can socialize, pursue hobbies or other recreation, or exercise—all of which will contribute to a sense of balance in your life.

Among these overlapping circles, you should try to find someone to serve as an intellectual partner: someone who can broaden your view and, in a supportive way, challenge your ideas, someone with whom you feel comfortable exchanging ideas and getting feedback. This person need not be a collaborator, or even someone in your research group, but his or her mind should mesh well with yours. The ideal is Watson and Crick, but less than ideal can still be useful. Spend informal time with your partner, over coffee or a beer, or hiking on weekends. The best conversations and ideas happen when you're relaxed.

COMMUNICATE INFORMALLY

As you develop your ideas, make a point of communicating them, at first informally and then more formally, to a gradually widening circle: first your labmates and intellectual partner, then your research supervisor, then a group meeting, then a departmental research seminar, then as a paper at a regional or national meeting. At each stage, you will get valuable feedback, constructive criticism, new ideas, and affirmation that you are on the right track in your research. Demonstrating to yourself and others that you can compete intellectually, that you can defend your ideas and your experiments, that you have a lot to contribute, that you belong in this community of scientists, will instill a growing confidence and pride in your results. Let others know about them, and relish their appreciation of a clever idea or a solid set of experiments.

PUBLISH

The final test of a scientist is publication in a peer-reviewed journal. The highest-ranked graduate students and postdocs, the ones with the most promising careers, are those who have published several major articles based on their dissertation or postdoctoral work. Seeing your work in print, knowing it has passed the scrutiny of experts in the field, can provide a powerful boost to your confidence and your motivation to do more. Preparing a piece of research for publication is a lot of work, but if you have done the preliminaries properly—developed your work systematically, kept a good notebook, and tried out your ideas in a variety of informal settings—you should have no trouble.

BE ENGAGED WITH YOUR DEPARTMENT

An important study by Barbara Lovitts of the reasons why students drop out of graduate school without completing their PhDs reveals that the most important reason is lack of engagement with others in their department. Be sure that you come to the lab, interact with the other researchers, and view your responsibilities as a teaching assistant as an opportunity for interaction. A department or graduate program is not just a place to get a professional credential; it's probably your most important psychological home (along with your family, if you have one) during these critical developmental years.

BECOME A PROFESSIONAL

Your goal in going to graduate school should be to become a member of an elite profession, distinguished by skill, knowledge, and accomplishment. At this early stage of your career, you've still got a distance to go, but there are many ways in which you can act like a professional, model your behavior on that of more senior professionals, and generally convince yourself that your goal is being achieved. As the excellent Web site maintained by the University of Michigan's Rackham Graduate School says, "Take Yourself Seriously. Make the transition from thinking of yourself as a bright student to seeing yourself as a potential colleague.

- "Attend departmental lectures and other activities.
- "Join professional associations and societies.
- "Attend conferences and use these opportunities to network with others.
- "Seek out opportunities to present your work (in your department or through outside conferences, publications, performances).
- "Attend teaching workshops and discipline-specific pedagogy classes."

GET RECOGNITION FROM SENIOR FIGURES

Once you've gotten some good results and have presented them at a meeting or in a paper, your work is likely to draw the attention of prominent senior researchers in your field. Nothing is more psychologically rewarding, more bolstering to your self-confidence, than being noticed in this way and realizing you've done something that others who have contributed to the field find important. As Csikszentmihalyi says:

At some point in their careers, potentially creative young people have to be recognized by an older member of the field. If this does not happen, it is likely that motivation will erode with time, and the younger person will not get the training and the opportunities necessary to make a contribution. The mentor's main role is to validate the identity of the younger person and to encourage them to continue working in the domain. The guidance of an older practitioner is important also because there are hundreds of ideas, contacts, and procedures that one will not read in books or hear in classes but that are essential to learn if one hopes to attract the attention and the approval of one's colleagues. Some of this information is substantive, some is more political, but all of it may be necessary if one's ideas are to be noticed as creative. (1996, 332–33)

BE ALERT TO JOB POSSIBILITIES

Fear that you won't get a job at the end of your training is a potent de-motivator. Do things that will inform you about a range of job possibilities and put you in touch with people who may be helpful. Network, go to career seminars, talk with people at meetings. Remember that even though you may not get your ideal job, there are lots of good jobs related to your professional training. Holders of PhDs in the sciences and engineering have about half the unemployment rate of the public at large.

Dealing with serious problems

Up to this point we've mainly considered ways to sustain your motivation when faced with the small, transient discouragements that are common in any prolonged, challenging activity. But sometimes more serious problems may threaten to derail your career aspirations.

FINANCIAL PROBLEMS

Most PhD candidates and postdocs in the sciences or engineering receive financial support through teaching assistantships, research assistantships, or fellowships. These are usually accompanied by fringe benefits in the form of tuition and fees (especially health insurance). So, while you may need to be frugal, you should have enough to live on. MS candidates, most common in engineering, may have to pay their own way (usually only for a couple of years) but will enjoy good job prospects when finished. If you do run into problems, perhaps due to health or family emergencies, your university probably has a student emergency loan fund. You may also be eligible for subsidized housing, food stamps, and state-covered health insurance, particularly if you have a family. Your institution's financial aid office should be able to guide you.

HARASSMENT AND DISCRIMINATION

Sexual harassment, or discrimination on the basis of race, ethnicity, gender, or disability, can be major deterrents to effective work. These are illegal, and nearly all universities have strict codes of conduct and university offices that will help you with such problems. Some states and many universities also bar discrimination on the basis of sexual orientation. Don't let such situations fester. If you are having problems with a faculty member (your advisor or someone else) or another student, and can't resolve the problem face to face, first talk to your department chair, director of graduate studies, or some other trusted administrator. You can also get help from university offices that deal with equal opportunity or affirmative action, disability services, grievances, dispute resolution, or other relevant subjects.

CHANGING LABS

Even with all the motivational aids we've discussed, you may reach a point where nothing seems to work and your whole graduate or postdoctoral plan no longer seems relevant. If a couple days of relaxation doesn't restore your enthusiasm, devote some time to asking yourself why. If you're having trouble in the later stages of a project, you may need to step back and rethink the approach. Maybe the analytical method just isn't sensitive enough, or there's some interfering substance, or the reaction is too fast or too slow for the method you're trying to use, or the variables have not been cleanly separated. Or maybe the concept is just plain wrong, and it's time to admit it and stop beating your head against the wall (or letting your research advisor beat your head against the wall).

All along, creativity and persistence and flexibility have been important. But sometimes creativity and persistence don't work, and you need the flexibility to consider another laboratory. If you're at an early stage of your graduate work, you may have concluded (perhaps subconsciously) that this is not the right project or the right lab for you. You don't find it intellectually engaging, or it's too hard or too trivial or too theoretical or too empirical. Or you don't like your research advisor or other people in the research group. (One way to avoid such problems, by allowing students to assess various options, is through lab rotations, a couple of months spent in each of three or four labs in the first year of graduate work. Rotations are common in the life sciences but not in physical sciences or engineering.) Whatever the reason, in these circumstances you should consider changing labs.

Within the first year or two the switch, made in consultation with the director of graduate studies (DGS) and/or the department head (and, of course, your major professor), should be relatively easy. If these things are done openly and honestly, they usually occur smoothly and without hard feelings. If, however, they are done surreptitiously, they can cause problems. Of course, you should discuss your concerns first with your research advisor, and see whether a new project might be both feasible and more appealing. You might well find that your advisor has the same sense of unease about how you're fitting into the research group and is supportive of some kind of change.

If you're further along in your graduate career, switching projects or laboratories will be harder but not impossible. Again, frank and open discussion with the concerned faculty members and DGS is the way to go. It's best, however, not to let things wait this long. If at all possible, pay attention to any vague feelings of unease early on, ask yourself why you feel that way, and figure out what to do about it.

Similar considerations hold for postdocs. Changing labs can be difficult if you're a postdoc employed on a PI's research grant, since you're beholden to that one investigator, who is relying on you to work on the project and who can influence your future career prospects. If you've come with "your own money" thanks to a fellowship, your freedom of movement is greater. And regardless of financing, unlike a graduate student you don't have to worry about transferring courses or credits or satisfying new degree requirements.

RETHINKING YOUR CAREER GOALS

If feelings of dissatisfaction persist despite all efforts to dispel them, more serious thought and action are necessary. First, recognize that the prolonged work and delayed gratification leading up to a PhD, and the three to five years of postdoctoral studies that may follow, are indeed difficult, although in most cases the rewards outweigh the difficulties. So you need to ask yourself again whether the PhD or an extended postdoctoral is really an important goal for you. If it is—if you really want to teach in a college or university or have a high-level job in a pharmaceutical or chemical company that requires a PhD and some postdoctoral experience—then there's no getting around it. You'll have to gut it out.

If you have doubts, however, that there will be a good job waiting at the end of the rainbow—an increasingly common doubt these days, when the number of PhDs on the market outpaces the number of employment opportunities in higher education—then perhaps you need to investigate other possibilities more imaginatively and thoroughly. In the sciences, especially, there are many good jobs in industry or government that can be satisfying alternatives to a teaching position in a highly ranked research university or liberal arts college. Going to law school and getting a specialization in intellectual property law is another possibility that more and more students and postdocs are investigating.

It's OK to not finish, if your goals or priorities have changed. You may find, for example, that the area you're committed to living in offers few job opportunities for a PhD in your specialty, or that family responsibilities have grown. As the Council of Graduate Students at the University of Minnesota has written (1999): "You are not your degree. Your degree is not you. Success (or failure) in graduate school has little, if anything, to do with your worth as a person. Graduate school is not worth risking your personal relationships or your health."

GET PROFESSIONAL HELP

If feelings of discouragement, low self-confidence, or low self-esteem persist, take advantage of the counseling services at your university. The professionals in these offices have had a lot of experience dealing with problems like yours. Don't try to go it alone; take advantage of their skills.

Take-home messages

- Keep a positive attitude.
- Be prepared for challenges as well as success in research.
- Prioritize your goals and streamline different tasks.
- Divide large tasks into segments that can later be reassembled.
- Maintain focus; avoid distractions while you are working.
- Take breaks and allow yourself to enjoy life away from the research environment.
- Get enough exercise and sleep. Maintain a healthy diet.

References and resources

Allen, David. 2002. *Getting Things Done: The Art of Stress-Free Productivity.* New York: Penguin. See also http://www.davidco.com/.

Csikszentmihalyi, Mihaly. 1991. *Flow: The Psychology of Optimal Experience.* New York: Harper & Row.

Csikszentmihalyi, Mihaly. 1996. *Creativity: Flow and the Psychology of Discovery and Invention.* New York: Harper Collins.

Lakein, Alan. 1973. *How to Get Control of Your Time and Your Life.* New York: New American Library.

Lovitts, Barbara E. 2001. *Leaving the Ivory Tower: The Causes and Consequences of Departure from Doctoral Study.* Lanham, MD: Rowman & Littlefield.

Parent, Elaine R., and Leslie R. Lewis. 2005. *The Academic Game: Psychological Strategies for Successfully Completing the Doctorate.* West Conshohocken, PA: Infinity Publishing.

Rackham Graduate School, University of Michigan. 2006. "How to Mentor Graduate Students: A Guide for Faculty at a Diverse University." http://www.rackham.umich.edu/StudentInfo/ Publications/FacultyMentoring/Fmentor.pdf.

15 MANAGING TIME

Time is irreplaceable

There are only twenty-four hours in the day, and most of us have many more than twenty-four hours' worth of things we'd like to do. While you're in graduate school, the most important thing you can do is to attend to graduate school matters: study for your courses and prelim exams, TA your courses, and do your dissertation research. If you're a postdoc, your top priorities will be to do research and scout out potential jobs. But you may also have family or personal responsibilities, hold down an outside job, and need time for exercise, sports, or other kinds of recreation. How do you fit it all in? This is where time management comes in. "A small investment in good time management can bring a large return in accomplishment and self-esteem" (Reif-Lehrer 1990). Not only will you get more done, but you'll feel more in control of your life. Planning and scheduling give you evidence that your time is being devoted to the things that matter most to you.

The portion of your time that is precisely scheduled for you has decreased continually, from high school to college, college to graduate school, and grad school to postdoc. You now have relatively few fixed courses or other responsibilities, but lots of things that you should do, so your ability to manage and schedule your own time becomes more important.

Good time management is not restrictive; rather, it leaves you more time to do things you want to do. Invest fifteen to twenty minutes a day in planning, and you'll save time. Time management involves two main things: deciding what you want to do, and making time to do them. In this chapter, we'll set out a systematic procedure for managing your time, see how these two aspects proceed in an iterative, ever more specific fashion, and give you some useful tips and tools.

Set your goals

The first step in effective time management is to define clear, realistic goals. This is harder than it sounds. You may have more goals than you can realistically accomplish. You may find that some of them won't matter very much a year from now. Others change as you change. Goals need to be revisited periodically.

It's a good idea to set goals with a long—but not too long—time horizon. For a beginning graduate student, winning the Nobel Prize or even getting a tenured faculty position is too far out, but getting a PhD in molecular biophysics and a good postdoctoral position or beginning professional job in the next five years is reasonable. For a postdoc, a three-year goal of doing research in nanotechnology applied to cancer therapy, which leads to published papers in top-ranking journals and a job in a leading research university or industrial laboratory, seems about right.

You should have personal as well as professional goals, to maintain balance and richness in your life. These might include developing loving, long-lasting relationships and friendships, hiking the Appalachian Trail, or becoming a passable saxophone player. Aim to do the things that seem really interesting and important to you. Your priorities may change, but if you eventually find that a seeming top priority is not for you, at least you'll have tried it out, rather than fretting wistfully about it. At the same time, have realistic expectations. Don't give yourself so many goals that you're bound to be disappointed or to short-change them all. Give yourself time for fun and relaxation, at work and away.

List the objectives you need to accomplish to reach your goals

Once you've set your goals, list the objectives you need to achieve in order to reach them. For example, if you're just beginning graduate school you will need to satisfactorily complete your program's course requirements, pass written and oral preliminary exams, fulfill any teaching requirements, carry out the research needed to write a dissertation, write the dissertation, and arrange for your next job or postdoctoral position. You will be aided in identifying these objectives by the orientation material you should receive when you begin your graduate studies, and by subsequent conversations with your director of graduate studies, thesis advisor, and other graduate students.

If you're an international student and English is not your first language, you'll also have to develop your English language skills to the level at which you can understand and participate in classes, work as a teaching assistant, and write for class assignments, written prelim exams, and journal articles leading to your dissertation.

If you're a postdoc, your objectives will include learning the background and techniques of the new area into which you've moved, perhaps writing a fellowship application for independent financial support, doing the research and writing the papers that you hope will land you a good job, and making the contacts that will lead to that job.

Don't forget your personal goals. If you hope to hike the Appalachian Trail, your objectives might include arranging time to do it, reading about various approaches, getting in shape, and obtaining suitable gear.

Establish a timeline for achieving your objectives

Some of the objectives you've identified will need to be achieved earlier than others. If you're beginning graduate school, you'll have to pass your courses and select a research advisor before you undertake the research or start looking for postdoctoral positions. If you're a beginning postdoc, you'll have to get tooled up and produce some results on your research project before you start looking for jobs.

When you enter graduate school, your program's requirements are likely to be mapped out in considerable detail, especially for the first two or three years, in a handbook and orientation sessions. A typical program might involve three full semesters of classes with a half load in semester four, working twenty hours a week as a teaching assistant one semester each in years two and three, a preliminary written exam in the first semester of year two and a preliminary oral (with a thesis plan and some preliminary work to report) in the first semester of year three.

The program may not explicitly set a timeline for doing your research and writing your dissertation. Many graduate schools have a limit of seven years after passing prelims—check what the rule is at your institution. In any case, you should set yourself an ambitious but feasible schedule; e.g., most of years three and four, and all of your summers beginning between years one and two, devoted to research, and writing up and defending your thesis during the first semester of year five. Devote some of that last semester, and the summer before, to thinking about your next job—postdoc or real job—and beginning the search so you'll have some place to go when your thesis is complete.

Your plans may differ, but this is a fairly typical pattern for graduate school in the sciences. Write these milestones down and refer to them, updating as necessary, at the end of each semester and summer session (or each quarter, if you're on that system) to prepare for the next term.

If you're a postdoc, you should establish a timetable for doing the reading that will bring you up to speed in your new area, getting trained in new techniques (which may involve taking a course or workshop), writing a fellowship application (find out what the application deadlines are), and getting started in the lab. We strongly urge that you plan to spend no longer than three years as a postdoc in a single lab (and no more than five years total), so a milestone for the beginning of year three should include the start of serious efforts to find a real job.

Break your objectives into smaller projects

Now consider the objectives for the next few months, and divide them into manageable pieces so you can track progress. You can't accomplish a major objective all at once. Take one step at a time, while keeping the long-range goal in view. View each significant step as a project in its own right.

For example, at the beginning of graduate school one of your prime objectives will be to choose a research laboratory. This might be divided into smaller projects: attend orientation lectures where faculty members present overviews of their research agendas; schedule office visits with those whose research programs seem most interesting; read a couple of recent journal articles from each interesting lab; talk with students in those labs, to get a clearer sense of what life there would be like; and take a series of research rotations, if they're offered by your program.

Your main objective as a graduate student or postdoc should be to do some significant, publishable research. This, too, can be broken into smaller, more manageable projects. Some of the beginning steps might be to become familiar with the literature in the area, learn the techniques you'll need to get started, practice those techniques by working on a system for which results are already known, order or prepare your experimental materials, and obtain or write the computer programs you'll need to analyze your data.

If your research has bogged down and your objective is to come up with some fresh ideas, here are some finite-sized projects that might help: do a computer search of the recent literature, read a couple of review articles in a related field, think about a new experimental system that might clarify the questions better, think about a new technique that might give useful new information, talk to an expert, go to a meeting to talk with others working on similar problems, send an e-mail inquiry to a newsgroup, and reread chapter 13, on creativity.

You'll also want to do some concrete thinking about your personal objectives and to translate them into projects. Does your spouse or significant other have things going on that should be taken into account? Do you need to arrange carpooling or learn about public transit schedules? If you have children (it's hard to do grad school with a family but not impossible), do you need to arrange day care or preschool? Do you need to have an outside job, and how will you fit it in? Do you have a plan to get regular exercise and recreation, to keep mind fresh and body fit? Decide what projects you need to carry out in order to achieve these objectives.

As you define each project, put it on a list. When you're done converting your objectives into projects, you'll have a to-do list that incorporates all of the things you need to get done in the next few months to realize your objectives. Your educational, professional, and personal goals will all be represented as achievable projects.

Schedule the projects on the list

Each project will fall into one of four categories with respect to scheduling.

- Regular, recurring events: e.g., classes Mondays, Wednesdays, and Fridays 8–10 a.m. and 1–2 p.m.; departmental seminar Wednesdays at 4 p.m.; group research meeting Thursdays at noon.
- Things that should happen regularly but aren't rigorously scheduled: study for classes, go to the gym twice a week, go on a weekend date.
- Things that have a deadline for which you need to prepare: midterm and end-of-term exams, visit home over Thanksgiving, term paper due the week before classes end, poster presentation at professional meeting in February.
- Projects, mostly associated with your research, that have no definite due date, but on which you must make some progress in the next few months in order to reach your longer-term goals: read the last two years' most important review articles and journal articles, do the next series of experiments, write the first draft of a manuscript, pick three labs in which you'd like to postdoc or three companies for which you'd like to work.

Treat the projects in the fourth category like those in the third and assign them deadlines, even if the dates are arbitrary.

ASSIGN PRIORITY TO EACH PROJECT

Each of these projects is important to achieving your goals, but some may be more important than others. Which absolutely need to be done by a given date? Which can wait? Which need to be done before something else can be done?

ESTIMATE TIME FOR EACH PROJECT

How long do you expect each project will take? Give yourself a margin of error (50 percent is a frequent suggestion) to allow for unexpected delays, emergencies, and underestimates. Break down projects that are still sizable into smaller ones, some of which may take no more than an hour or so. Give yourself lots of lead time for important projects with deadlines, so that you'll have some flexibility if an emergency arises.

SCHEDULE THE TERM

Now take out your calendar and write down the due dates of your projects. Begin with a semester or several-month view and enter the timelines for all your long-term priorities and recurring tasks. If possible, schedule ongoing but noncompulsory tasks for the same time each day or the same day each week, so that you'll get used to the idea of, for example, working on the big machine Mondays 8–11 a.m. and scanning the literature Thursdays 3–5 p.m. This will eliminate one common drain on your psychic energy, wondering what you should do next. You'll also want to schedule as far ahead as possible any important activities that involve other people—your thesis committee's annual advisory meeting or regular meetings with your advisor, for example.

SCHEDULE THE WEEK

The next step is to do some more detailed planning for the coming week. On Sunday evening, or some other convenient and regular time, look through your calendar. Your regularly scheduled commitments should already be there, and you can add or subtract appointments that have come up or been canceled. Then look over your list of unscheduled priorities, add new ideas, and ask yourself whether you're committed to spending at least five minutes on each during the next week. If you don't expect to get to an item this week—you're just too busy, or the person you need to talk with is out of town, or the reagents for that experiment won't be in until next Tuesday—cross it off the list for the week, but keep it on your longer-term list.

SCHEDULE THE DAY

Finally, move one or two of the unscheduled things on your weekly list into open blocks of time on Monday and perhaps Tuesday. If you finish them, go back to the weekly list to bring up more items. If you don't finish them, move them to the next day. By the end of the week, you'll usually have completed the key things on your weekly list. The important thing is to have a plan for your day. Otherwise you'll end up doing just whatever comes along.

Try to do at least one substantial, significant, but discretionary thing every day: start the next experiment, read a key paper that just came out, order reagents for the next experiment, work up your data from the last series of experiments. Most of these require significant blocks of time, and it's easy to fritter away your uninterrupted time with trivial distractions unless you schedule your day and keep to the schedule. Accomplishing even one significant part of one objective each day will make you feel better about your progress and help maintain your motivation.

SCHEDULE THINGS AT THE BEST TIME TO DO THEM

Another aspect of good time management is scheduling things when you're most fit to do them. Many people have the most energy and the clearest head early in the day. If you're one of them, try to do your most creative work—that which requires the hardest thinking and concentration—in the morning. It may be best to come in quite early, before many other people are around to break your concentration. In the afternoon you may hit a slump. Schedule routine tasks, or physical things that don't take a lot of mental energy, during that time. Going for a jog or a workout at the gym in the early afternoon might perk you up. On the other hand, if you're like many graduate students and postdocs you don't get started until late in the day, then work late into the evening. That's OK, but again try to do the most important and difficult things when you're mentally most alert and least likely to be distracted by other people.

Experimental work often requires lengthy blocks of time and thus demands thoughtful scheduling. It's often best to start experiments early in the morning. If an experiment involves extended periods of waiting while things incubate, try to make the downtime coincide with seminars or meetings when you're already expecting to be away for an hour or so. And if you can save several hours the next day by staying an hour late and starting an experiment tonight, do so.

BE FLEXIBLE

Tasks are never fully predictable. Give yourself an hour or two of flex time each day for unexpected events. Your estimates of time will become more realistic as you gain experience, but at any given moment, a thoughtfully prepared schedule is the best you can do. It will give you direction, conserve psychic energy, and give you confidence that you're proceeding toward your goals.

If things run seriously astray, don't waste time berating yourself or brooding over what is beyond help. Try to get back on schedule tomorrow, and think about why things didn't work today and what you might do in the future to avoid a recurrence.

Time-management tips

Make steady progress. Give yourself the satisfaction of making steady progress by working in good-size chunks of time. Starting and stopping a lot is wasteful. Put in consistent hours on research.

Know what you're going to do next. Know the next action required to move forward on each objective. For example, analyze your data at the end of the day, so that the next day's work is clear. Writers sometimes leave the last sentence of a paragraph untyped, so they can start the next day by finishing what they intended to write, then use that momentum to move to the next paragraph. Or they retype the last finished paragraph, and then go on. This can be useful in making progress on a big writing project like a journal article or a thesis.

Do two things at once. Plan to have more than one thing happening at the same time. Most lab research entails periods of intense activity requiring close concentration, but these may be interspersed with slack periods during which you could plot your most recent data, update your notebook, browse a journal, or set up your next experiment to run overnight.

Do it right the first time. As the proverbs say, "Haste makes waste" and "A stitch in time saves nine." Rushed experiments will likely need to be done again. Give yourself enough time to do them right. Be sure your tools are in good order, your instruments are properly and regularly calibrated, and your reagents are functional. Even in preliminary experiments, use good technique; don't waste time repeating work unnecessarily.

Think about your results as you proceed. Graph, analyze, and think about your data as you go along. Don't just tabulate numbers. Graph them to spot trends or undue scatter. Try to have a clear idea of how an experiment should behave, so you can recognize a malfunction or a genuinely new result. Be as quantitative as you can in your expectations. If you're doing experiments, compare them with theory. If you're doing theory, test it against an experiment.

Keep good records in your notebook. Have a system for handling supporting data and calculations that are not entered in your notebook. Keep a clear record of your references and sources, for use in paper and thesis writing.

Plan your work so that each experiment, calculation, or piece of reading makes a definite contribution to your progress. Each experiment should yield a point on a graph that will appear in a publication. Alternate experiments, analysis, and reading. Don't wait until the end to learn what's in the literature or to start writing up results.

Change the stimulus. If you're running out of steam doing one thing, do another that also contributes to your goal. Many scientists love the lab so much that they find it pleasant and natural to work fourteen-hour days, staying until the wee hours of the morning. But those people are not necessarily more productive, and they may waste time instead of planning carefully and getting things done in a reasonable workday. Don't be intimidated by a workaholic culture, feeling bad if you keep reasonable hours. You'll ultimately be judged by your productivity.

Don't waste time. Concentrate and focus; don't allow yourself to get distracted from your schedule. Be able to say "No" or "I'm busy now; let's talk about that later." Avoid computer games, idly surfing the Web, or checking your e-mail every five minutes. If you've worked too long and need a break, some physical exercise will probably be more useful.

Reading and talking to people about research are valuable, but don't overdo them or use them as excuses for not moving ahead steadily with your own experimental and analytical work. The same goes for service in your research group, department, or university: Do your share, but avoid trivia and try to spend most of your service time on the things that really matter to you.

Avoid clutter that causes you to waste time looking for the things you need. Have a good filing system and adequate shelf space. Arrange your computer files in a way that makes them easy to find.

If something is bothering and distracting you, complain to someone who can do something about it.

Handle paper and e-mail efficiently. You'll always have a lot of reading and e-mail to handle. Act promptly on the things that require immediate attention, throw away the stuff that's not worth attending to, and save the useful but not urgent mail and reading for downtime.

Capture your ideas. Don't let ideas slip away; write them down, preferably in your notebook. Some you can act on immediately. Others can be put on your project list or given to someone else and scheduled for follow-up. If possible, put them with related ideas in a file or computer database.

Deal efficiently with course work. At the beginning of your graduate career, you'll still be taking classes—fewer than as an undergraduate, but probably some involving more difficult and independent work. Since you got into grad school, you probably got good grades and had reasonably good study skills in college. Don't forget to use them now. The main ones are

- Make a semester and weekly schedule.
- Prepare for class; review your notes before and after each session.
- Work on material steadily. Don't cram.
- Make a note of term paper assignments and get started early.
- Study in a concentrated manner for moderate lengths of time (about an hour), then take a five- or ten-minute break.
- Ask questions as they arise.
- Read by skimming first, summarizing after.

Use available resources to solve personal problems. Don't try to be totally self-reliant, or be ashamed to take advantage of resources you may need. If you're being distracted by financial problems, talk to the university's financial aid office, arrange a loan, or get food stamps. If you have child-care problems and there's a relative nearby, use him or her when feasible (Felder 2004). If you're feeling ill or depressed, consult a university clinic.

Monitor your time planning. Your schedule is not written in stone. You have to be flexible enough to adapt to new circumstances, to realize that some things will take longer and some less time than you anticipated. Perhaps you didn't allow enough time at the end of the day to write up your results in your notebook. Perhaps the journal article for which you'd allowed two hours only took an hour to read, or a student who'd made an appointment didn't show up, and you didn't know what to do with the free time. Try to analyze what needs to change, and modify your plan. After a few weeks, you'll probably find the right rhythm. Then when a new semester comes around, with a new set of responsibilities, you'll have to rework your plan. The point is to be as mindful of time planning as of anything else in your life, and to work on getting it right. But don't be too hard on yourself: no one gets it right all the time.

Overcome procrastination

Everyone procrastinates. It's not surprising that we put off things that are unpleasant and give preference to things that are less productive but more fun. But sometimes we also put off things that are not overtly unpleasant and can't figure out why. Usually there's a nagging little hint just below our conscious awareness: too many things to do, no clear direction, the project is too big, there's not enough information, we fear inadequacy or failure. Here are some tactics to overcome procrastination.

Review your goals, objectives, and projects. If you have lots of things to do and no clear sense about what to do when, review your list of goals, objectives, and projects and remind yourself how each task will bring you nearer to your goals and objectives. Be realistic in your goals, recognize that you won't immediately be able to get everything done that you'd like to, and include breaks and recreation in your planning to avoid burnout.

Reassess the importance of your project. Maybe your procrastination is based on the intuition that the project you've chosen to work on is not really the most important thing you could be doing right now. (That new set of experiments might be made irrelevant by a just-published paper, or you sense you need to do more controls.) Make sure your project is really the best one to do right now.

Break a big project into smaller ones. Sometimes a project is just too big and complex to seem feasible. Try to break it down into discrete, definite small pieces. Determine the order in which these small things need to be done, then pick an early one and do it. For example, you can't write a journal article all at once, so identify the subtasks: get the instructions for authors from the journal or its Web site, write a sample title page, assemble the references, draw the graphs, write the captions, assemble the tables, write the materials and methods section, and so on. Doing these small tasks one at a time will soon result in a complete manuscript. Or if you need to study five hours for a final exam, plan to spend one hour on the material each of the five days before the exam.

Do an easy next action. If slicing the project into hour-size tasks still leaves you paralyzed, try an immediate next action. Label a set of file folders, set up a template on your computer for data acquisition, make some notes or a list. Find a small task that can be done in five minutes or less to get you started and pulled into the project.

Get more information. Sometimes your subconscious mind realizes that you don't have enough information to do the job adequately and sets up a block. If this is the problem, try reading, making notes, asking questions, and gathering the information needed to proceed.

Don't expect immediate perfection. Sometimes procrastination is not caused by laziness or distraction, but by your insistence on perfection. You think to yourself, "I can't possibly do this as well as I'd like to, so I won't be able to do it at all." Don't give in to perfectionism. Nobody's perfect, and every outstanding piece of research that you've admired is the result of many revisions. Get started, however imperfectly, and get involved with the task. Progress, revision, and improvement will come.

Identify and confront your fears. Sometimes procrastination is due to conscious or unconscious fears: This is too hard for me; I'm in over my head; even if I do this I won't get the job I want; somebody is likely to scoop me; my advisor doesn't like me; my boyfriend is angry at me because I spend too much time at the lab and not enough with him. Get these fears out in the open. Make a list, write them down, free-associate. Then ask yourself: Is this really plausible? If the outcome I fear were realized, would it be that bad? What are the likely consequences if I give in to this paralysis and don't get the work done? Which things can I actually change Should I talk to someone—my advisor or mentor or the counseling center—about this to get perspective and advice?

Take responsibility and move on. Admit to yourself that you're goofing off or escaping from the tasks that need to be done. Everybody does it, so you're not a bad person. Don't add the further fear that you're a motivational misfit, unable to keep on track with your important goals. Realize that you've chosen to goof off and now you can equally well choose to get back on track.

Get professional help. If all else fails, seek professional help. Your university counseling center is very familiar with student procrastination problems and can suggest ways for you to break out of a block.

Take-home messages
- Define clear and realistic goals, short- and long-term, personal and professional.
- Prioritize your goals.

- Make a list of the smaller steps you need to complete to reach each goal.
- Establish a reasonable timeline for achieving your objectives.
- Plan daily, weekly, and monthly schedules, but allow some flexibility to accommodate the unexpected.
- Learn to multitask and to juggle multiple responsibilities.
- Periodically review your goals, objectives, and accomplishments.
- Maintain a focus on finishing tasks in a timely fashion. This will allow you time for your personal needs and goals.

References and resources

Allen, David. 2002. *Getting Things Done: The Art of Stress-Free Productivity.* New York: Penguin. See also http://www.davidco.com/.

Felder, Takita. 2004. "Nothing Is Impossible to a Willing Heart." http://nextwave.sciencemag.org/cgi/content/full/2004/05/19/4.

Hopkin, Karen. 1998. "Balancing Lab and Life: Could Science Ever Be 9-to-5?" *The Scientist* 12, no. 6 (March 16): 11–12.

Lakein, Alan. 1973. *How to Get Control of Your Time and Your Life.* New York: Signet.

Reif-Lehrer, Liane. 1990. "Suggestions for Saving Your Time and Keeping Your Cool." *Scientist* 4, no. 4 (February 19): 24.

16 FINDING AND MANAGING INFORMATION

The value of information

Information, like laboratory apparatus or computer, is a tool for research. It helps you develop ideas for new research, devise ways to pursue your current research more effectively, and avoid repeating work that has already been done. Researchers must seek information actively and purposefully, not wait idly for it to drift into view. Once obtained, information should not be passively absorbed; it must be examined, connected, and used.

Science is a cumulative activity: the information produced by those who have worked before you allows you to go further. If you are not aware of what has been written in the books and journals of your specialty, you may unnecessarily duplicate work or end up accused of overlooking or not citing work you should have known about. At best you will have wasted your time; at worst you may alienate researchers whose work you have ignored, or even be accused of plagiarism.

Getting relevant information is essential to effective research, and overlooking published information can be hazardous to your career, but finding and managing information is hard. We're overwhelmed with information. More books and journals are published, even in the most specialized areas, than anyone can possibly read. The scientific literature is doubling every five to ten years, and although much of this literature is redundant, it is not obvious how to separate the wheat from the chaff. Written information is not the only problem. At any large research university it's virtually impossible to find time to listen to all the seminars and lectures that might be pertinent to your work; and each specialty has more national and international conferences and symposia, at which the newest results are reported, than one could possibly find the time or travel budget to attend. Given this flood of information, identifying the truly important advances and finding the particular technical developments that may be key to one's work is becoming increasingly difficult.

This chapter is about strategies for finding and managing the information you need. It discusses the different kinds of information you need at different stages of a research project, the various written sources in which such information can be found, and the use of libraries and computer databases. It describes ways to take notes systematically and to document your searches, and how to read efficiently for various purposes. And it offers suggestions on getting information from seminars, meetings, and personal contacts.

It's rare that you need a piece of information only once. Say you find a technique for performing an assay: you will need the reference again when you write the paper and yet again when you teach another member of the lab about the technique. Thus, this chapter also discusses how to develop a system for storing and retrieving information, so that you can find things that you know you have and be reminded of things you've forgotten.

It's important to develop a system that helps you cope with information, but don't overdo it. Information is a tool that enables you to be more effective and creative, not an end in itself. Many days at the lab bench can sometimes be saved by a few hours in the library. Don't, however, spend too much time creating elaborate filing systems, or so much time reading the literature and going to seminars that you never do your own work. It has been said that the physicist Enrico Fermi spent two-thirds of his time on his own research and one-third learning what others were doing. That may be as good a guide as any.

Different kinds of information for different purposes

We are deluged with information but challenged to find the information we need. The huge number of publications in specialized journals (now well over a million articles per year), books, and conference proceedings makes it virtually impossible for experts to keep up with their fields, let alone for neophytes to learn the important new developments in a different field. The Web has given us more convenient access but challenges us to pick out the meaningful items from the background noise.

You will need different kinds of information at different times and for different purposes. The questions it answers can be broad (What is the current status of the field of gene therapy?) or narrow (How do I isolate this enzyme from pigeon livers?) Both kinds are important, but they serve different purposes. Depending on your objective and the stage of your research, you may need an introductory survey of an area, an intensive review, specific technical information, or an update on your field of specialization. Be clear about what kind of information you need, and seek out the most appropriate sources for each.

There are essentially three ways of getting necessary information: reading the literature in your field, attending presentations by active researchers, and talking with people who know useful things. We'll discuss effective strategies for each in the next three sections.

Reading the literature

READING WITH A PURPOSE

Your main purpose in reading, and in gathering information in other ways, is to build up expertise—a structure of self-consistent knowledge and a set of important, unanswered questions—in your area of research. In the course of a project, a general idea gets continually modified, expanded, and contracted. The scope is never clear at the beginning.

Read with questions in mind. When you are starting a project, your main question should be What has been done? (and implicitly, What has not been done, or has been done incorrectly or misinterpreted?). By surveying the literature to answer this question, you can build on the results of others and avoid duplication. This is also the time to begin assembling a bibliography for the background chapter of your dissertation, for the introductory sections of papers you will write, and perhaps for grant or fellowship proposals. Once you are immersed in a project, your questions might be

- What has been published recently that affects my project?
- How can I solve a particular technical problem?
- How does this article help me interpret my data?
- Does the paper give me ideas for future work?

Sometimes you will scan the literature not primarily for material relevant to your research, but rather to broaden your awareness of current developments in science (What has happened recently that is interesting?). You should of course regularly browse a couple of the major journals in your field. But you may also wish to peruse such journals as *Science, Nature,* and *Scientific American,* which cover a wide range of the most important current discoveries, often accompanied by commentary that makes the articles more accessible to the nonspecialist. Spending some time broadening your general scientific knowledge will make you more alert, better able to understand what others are doing, and more receptive to the occasional opportunity to use a development in another field to move your own work in a productive new direction.

For an introductory survey of an area, the best starting point will be a technical encyclopedia (if not too out of date) or a current textbook. If, for example, you know little about photosynthesis but want to learn, a general biology text will lay out the basics. A biochemistry or plant physiology textbook might be an appropriate next step. Depending on how wide or deep you need to go, a monograph may be next—but be sure it is relatively up to date. By this stage you should have identified the subtopics of greatest pertinence to your research. The next step is the review article. There is an increasing number of review series (Annual Reviews of . . . , Trends in . . .), in which leaders in various subfields identify the most important recent developments and put them in perspective. These reviews are most valuable if they are critical and convey a definite point of view, rather than being simply compendiums of recent results. Be alert, however, to the potential for biases that do not address opposing points of view. You should never cite a reference listed in a review article without actually reading the original paper.

Intensive literature surveys are undertaken for a purpose such as writing a grant, the background of a paper, or a review article. You will usually want a fairly complete listing of all references available, though once you start reading you will find that many references show up repeatedly and many are trivial, so can be dealt with quickly. Reviews are the best place to start. A couple of key review articles, updated by surveying the most recent issues of the major journals, will generally provide what you need. As discussed in detail later in this section, computer searching of bibliographic databases has revolutionized the process of looking for articles on particular topics or combinations of topics. Titles, authors, abstracts, and bibliographic information for many articles can be obtained in just a few minutes. Often the main problem is to narrow the search so that you retrieve just the pertinent sources. Citation indexes offer another useful strategy. Each field has certain key papers that most subsequent articles are likely to cite. Science Citation Index lists all papers that have cited a given paper, thus tracing threads through the literature of a field.

To proceed effectively with research, you will often need information on specific technical data and procedures. Questions here are of the following sort: What is the boiling point of this compound? Where can I find a computer program to integrate this set of differential equations? How do I perform this analysis? This kind of information is often not easily accessed through computerized searches. Handbooks and collections of tabulated data and methods must be consulted. The reference section of a university library will generally have a collection of basic reference sources, dictionaries, encyclopedias, and handbooks for the topics held in the collection of that library. If you think information probably has been published, but you can't find it rapidly, ask a librarian. They're highly trained in information retrieval and are delighted to be consulted. Alternatively, asking an expert (discussed later in this chapter) may be the most efficient approach.

READING EFFECTIVELY

Read systematically. Reading should be conducted with discipline, keeping to a definite schedule. Follow the journals that publish the best papers in the field. Learn which groups are doing the best work, and concentrate on reading their papers. Establish computer search profiles for key topics to get regular updates.

Read actively. Good readers think and ask questions as they read. Anticipate where the author is going. Keep asking: What is the main idea being presented? How does it relate to what I already know? Why am I reading this particular piece, and how can it help me in my work? Is it answering a question that I have? Does it reinforce, contradict, or clarify a theory I have? Does it give me a particular technical answer? Even though I'm just browsing for intellectual breadth, how can I connect this to what I am working on?

Read efficiently. Read at a time when you are not too distracted and can pay attention. You may find it best to read first thing in the morning, when you're more receptive to other people's thoughts and your mind is less full of your own. Learn to skim, and concentrate on picking up main ideas. Don't read journal articles straight through; focus on getting the meat quickly:

- *Title.* Should tell you what the article is about; might signal that you don't need to read further.
- *Summary (abstract).* Is there something new that's worth remembering?
- *Introduction.* Poses the problem and its background; explains inadequacies of prior efforts. This section may be most useful to beginners.
- *Discussion.* Needs to be read skeptically. Look here for the conclusion, particularly if abstract is not clearly written.
- *Results.* Skip this unless you're seriously interested in evaluating the paper. Look at figures and tables first, then at text to see whether it agrees with your impressions of the data.
- *Materials and methods.* Useful only if working in the field, to pick up useful hints or new methods.

Focus on graphs and tables. In technical articles, a figure or a table has the same status as a paragraph and often conveys more information. Some scientists feel you should be able to grasp the entire message of an experimental paper by looking at just the figures and their captions. Some questions to bear in mind:

- What are the units?
- What is the baseline? If it's not zero, small effects may be exaggerated.
- Is it a logarithmic plot? Large variations can be hidden in a log plot.
- Are there error bars? Should there be?
- If there's a comparison of two sets of conditions, or of theory and experiment, do the precision of data, and sensitivity of the theory to key parameters, really allow a strong conclusion?

SEARCHING SYSTEMATICALLY

Searching for and acquiring information is a never-ending task. You should do periodic, perhaps weekly, sweeps through the literature—especially the most important journals for your field—to see what's new. Frequently you will have to search for particular information, such as how a particular assay is performed. And if you're beginning a new phase of your project, you should do a literature search on that phase, even if you've done a more general one earlier, to make sure that you picked up all the relevant material. Now that you're more knowledgeable in the area, you'll be able to construct a better search strategy, one that misses less (fewer false negatives) and comes up with more pertinent references (fewer false positives).

As a member of a research group, you shouldn't have to be entirely self-reliant in this regard. Subscriptions to a few of the main journals and to on-line resources that are updated weekly, such as *Current Contents* or *Medline*, can be shared; periodic Selective Dissemination of Information (SDI) updates passed around the group; and a customized reference database made available on networked lab computers. Group members can be assigned responsibility for keeping up with a few journals each and for reporting on important articles at weekly group meetings or journal clubs.

ONLINE SEARCHES

With the explosive growth of the scientific literature, locating the information you need can seem like trying to find a needle in a haystack. Fortunately, computer indexing and online databases have at least partly solved the problem. If you're in a university or private-sector research lab, you will generally have access to Medline (PubMed), Current Contents, and a variety of specialized databases. It's easy to run searches and find all articles published, for example, by John Smith or Roger Jones on the effects of aspirin on fever in the past two years. Abstracts are generally available, and if the article was published relatively recently (typically since the mid-1990s) the full text will be accessible as a PDF or HTML file for many journals. Thus in a few minutes, without leaving your desk at work or at home (especially if you have a fast modem, DSL, or cable connection to the Internet) you can do a large amount of library work and in many cases even get the full papers.

Of course, such a specific search is not what you'd normally begin with. You might start with a search of PubMed for articles published on aspirin and fever over the past five years. Supppose that turns up 124 articles. If you then narrowed your search to review articles in English, that might give you 35. After looking over a few of these, you might decide to concentrate on mechanism of action, so you add "mechanism" to the search terms. You find that Roger Jones (a fictitious name), who's at a well-regarded university, has quite a few articles, and you conclude he's probably an expert in the field. So you search for all the articles published by Roger Jones in the past five years. In just a few minutes, then, you can locate a number of key review articles and get an idea of the leading researchers in the field, and you're on your way to a useful overview of the area in which you'll be doing your research. Of course, you still need to read those papers and take notes on them, but it's easier than ever before to take the first steps. And once you've found a particularly relevant article, you can look at the terms in its descriptor list, then revise your search strategy to include them.

If you know that a paper written several years ago has been influential, the ISI Science Citation Index (Web of Science) is a useful way of finding subsequent articles that have cited it and are therefore presumably concerned with the topic it covered. Your university library may have an online subscription that you can access from your desktop, or you can search in the library.

There are many advantages of online searching:

- Hundreds of thousands, if not millions, of items are available every day, around the clock, often as full text. Few libraries can claim comparable holdings.

- If full text is not immediately available, it can generally be ordered.
- Documents not available at many university libraries—patents, dissertations, conference reports, government documents— are accessible.
- Hundreds of items can be retrieved rapidly, at minimal cost in terms of person-hours.
- Complex searches using Boolean operators can be designed and carried out readily. (For example, if you know that a particular author published an article in 2005 on a particular topic, you can find it easily by searching for author AND topic AND PY=2005.) Or one can browse based on curiosity.
- Output can often be formatted to download directly into a reference database.
- Personal interest profiles can be used to order regularly updated searches.

There are also disadvantages:

- If full text is not available, there's a danger of deluding yourself that since you have the reference and abstract, you know what's in the article.
- Online searching may be costly in terms of access and retrieval charges. However, this is no longer the case in most university environments.
- Unless a search is carefully designed, it can retrieve either too many or too few items.
- The possibility of serendipity through browsing may be lessened.
- Online searching needs some training (though so does regular library use, and a basic facility can be achieved quickly).
- Online databases usually start in late 1960s, so you may not have access to valuable earlier literature.
- Some licensing agreements between publishers and libraries delay availability of the full text online for six to twelve months, though you can still access abstracts.
- You can become "literature-happy," collecting and filing references without actually reading or thinking about them.

Be careful not to plagiarize inadvertently when using online sources. It's too easy to copy abstracts or other text from a computerized literature search and paste them into a database or into a set of notes in your word processor, and then to incorporate them verbatim into an article, thesis chapter, or grant proposal. You should always reformulate such information and express it in your own words; or, if you need to quote, use quotation marks. In either case, cite the reference.

GETTING ARTICLES AND BOOKS

For most books and for journals not available online, you'll need an actual physical copy. Don't despair if the library doesn't have what you need, either because it's been checked out or because it's not in the collection (increasingly common in these days of rapidly increasing journal prices and static library budgets). Your librarian can usually arrange an interlibrary loan.

Going to lectures and conferences

Another way of getting information is to listen to scientists talk about their research. There are two main venues for this: lectures by visiting speakers at your institution and professional meetings.

You should attend all of the weekly seminars organized by your own department or program, even if they are unrelated to your research interests, and pick and choose among those in other departments. You will broaden your knowledge of your discipline, you will often get information that helps you interpret your data, and you may well get ideas for new projects. At a departmental seminar, you'll get an introduction to an active research problem, the latest results from a leading lab—most of them not yet published—and an opportunity to question the speaker about points that puzzle you and to try out your own speculations. You can prepare by reading some of the speaker's most recent papers before the lecture. There are often times scheduled for students and postdocs to meet with visitors (and these opportunities are sometimes underused, so take advantage of them). You might also ask your faculty mentor to arrange some time, speak with the visitor before or after the talk when people are gathered for refreshments, or join a lunch session.

Your department or program probably has a journal club, in which students and postdocs present articles from the current literature. Participation in a journal club is often a requirement of a graduate program. Although the speaker (most likely one of your departmental colleagues) may not be as expert as the author of the article, such presentations can both familiarize you with the current literature and (when it's your turn to present) give you practice in speaking about research results.

At professional meetings you'll find an abundance of people, from very senior scientists to students like yourself, working in the same or related areas. By visiting posters, attending talks, and chatting with people in the corridors, you can pick up an amazing amount of information. The challenge is to overcome natural shyness and actually make contact with people, and not to be overwhelmed by the sheer mass of information available. Conference programs, often available on-line or in a searchable CD-ROM format, can help you prepare your daily itinerary. If you make careful notes ahead of time on which posters and talks you want to see, and which people you'd like to talk with, a conference with a focus in your area can be an enormous boon. You will come back after three to five days exhausted but full of ideas and up-to-date information that you could not get in any other way. You may also be able to follow up contacts by e-mail. See chapter 18 for more details on going to scientific meetings.

Asking an expert

Sometimes the most efficient way to find a particular piece of information, or to find out whether a project has already been carried out (perhaps it failed and was never published), is to ask an expert. If you need to know how to purify a particular enzyme, or what's the latest good review article on mating factors in yeast, there's probably someone in your department or elsewhere on campus who can tell you. If you're trying to use a procedure that was published in a recent paper, and it isn't working, e-mail, call, or write to the author and ask for more detail. A five-minute personal exchange can sometimes save hours or weeks of work.

Remember, however, that even experts have gaps in their knowledge and prejudices about who are the best workers and what the most relevant results. You should respectfully accept their information and advice but, particularly as you grow more knowledgeable, supplement it through further searching.

Sometimes you will come across useful information by accident, in casual conversation over coffee or while using equipment in the neighboring lab. Not all "experts" need be world famous; the most helpful person may sometimes be a fellow postdoc or student just down the hall.

Managing information

Once you've gathered information from reading, searching databases, talking to people, and going to meetings, how can you keep track of it and use it effectively? It's one thing to find information, it's another thing to make it your own—to integrate it into your stock of knowledge so that you can locate it readily when you need it. Most of us are lazy and use information we can find easily; we won't bother to search for information—even information that could save us lots of time—if it's not readily accessible. Thus to the adage "Two months in the laboratory can save you a week in the library" we would add "An hour spent looking for an article you once read can save you the five minutes it would take to file it properly in the first place."

Your use of information will be made much more efficient if you develop an effective system that enables you to lay your hands (or more precisely, your mind) on the information that you went to so much trouble to locate in the first place.

There are three components to an effective information management system: assimilating the information mentally, taking notes, and filing.

ASSIMILATE THE INFORMATION

It's easy to persuade yourself that just because you have scanned titles and abstracts, photocopied or downloaded articles, and entered them into your filing system, you have actually assimilated the information they contain. It's crucial to make real intellectual contact with the material, to store it in your mind as well as in a database. As the molecular biologist Sydney Brenner once said, "Memox, don't Xerox."

It is impossible to read everything that should be read. However, it is not impossible to do a computer search of key topics each week, come up with twenty articles, put them in your reference file and database, mark seven that are particularly interesting or important, and read one each day. That's 365 important and interesting articles per year, at one-half to one hour each: a manageable but cumulatively huge input of information.

TAKE NOTES

Taking notes is an important way to assimilate new information, to select those aspects of an article, lecture, or conversation that are particularly germane to your interests. Note taking should be thoughtful, not a mechanical recording activity, and it should include your own thoughts and reactions as well as recording what you've read or heard. You may want to take notes on specific experimental conditions, main conclusions, relationship to other papers you have read, ideas for future directions, and your evaluation of the authors' data interpretation. Develop a consistent set of abbreviations.

Paraphrase rather than copying unless a direct quote seems called for, in which case use quotation marks. Write down the bibliographic details or source, so you can trace the information back to the original. Put the notes in your reference filing system and/or in your notebook.

FILE SYSTEMATICALLY

Physical filing. It's too easy, in the press of other business, to let papers pile up unfiled. Then when you want them to prepare a talk or verify a reference for a manuscript, you waste time rooting through stacks of disorderly paper. Try to set aside a regular time for filing and entering the references in your index system. The end of the week, when your concentration and energy are too low for more creative endeavors, may be a good time. You should have a "to be filed" box or folder, into which you put things each day. Once a week or so, unload it. Make sure you throw away things that turn out not to be interesting or worth keeping.

You will want to file not just reprints of journal articles, but also books, slides, overhead transparencies, PowerPoint presentations, data sheets that don't fit in your notebook, disks or CD-ROMs with archived data, research methods, maps, circuit diagrams—whatever you use in your research life. Also your own notes and ideas (i.e., the stuff in your lab notebook), and notes taken at lectures and conferences. Try to make connections between all the information in your intellectual life. Since these items have such different physical forms, an indexing and retrieval system will be needed.

Computer reference database. Until the early 1980s, references were commonly kept on index cards. Now one can use a computer reference database such as RefWorks, EndNote, or Bookends, where entries can be searched by topic, keyword, author, year, or other terms. The current versions allow you to load the results of online searches directly into the database. These programs are widely available, are not expensive relative to their value, run on popular personal computer platforms, and are among the most valuable tools of research.

Such databases can be valuable resources for an entire research laboratory. If the faculty mentor, students, and postdocs all contribute in a systematic and coordinated fashion, a customized database with thousands of references to a particular part of the scientific literature can be accumulated and accessed from networked lab computers. Such a focused database lends itself to much faster and more efficient searching than a more general online resource.

The value of such a database becomes even more apparent when one has to assemble a bibliography for a manuscript, dissertation, or grant proposal. As a manuscript is being developed, pertinent references from the master database can be copied into a new file. References subsequently added to the new file can then be added to the master file or kept separate, depending on their relevance to the lab's overall objectives. These database programs can also be linked to most common word processing programs, allowing specially formatted reference entries to be inserted in the body of the manuscript. When the manuscript is completed, the in-text citations can be put into the proper format for the journal and the bibliography generated automatically. This will work only if all users of the database are meticulous about entering data in the correct fields. You should never submit a manuscript without first reading over the complete text, including all references. New references can be easily inserted without renumbering, and the paper may be readily reformatted for submission to a different journal. Thus one of the most tedious tasks in manuscript preparation is made almost trivially simple.

It's important to keep track of the informal bits of information that you get in listening to lectures, going to meetings, and talking with people. You can enter them into either your master database or a separate one. Add keywords so the records will turn up in a search. Recording references as you go along will be invaluable as you develop the material more fully in notebook entries or a manuscript.

REVIEW FILES PERIODICALLY

Rooting through your own files periodically is like browsing in a library assembled by your twin. There's a lot of interesting stuff there, much of which you've probably forgotten, and you may make serendipitous connections between the papers you stumble upon, or between old and current interests. This can be highly productive. Though your interests may have changed, they are unlikely to be completely different. Perhaps you now see things in new arrangements, organized around new emphases. This is an opportunity to bring the filing categories back into alignment with your mental categories.

Take-home messages

- Information is a tool that enables you to develop your own creative ideas, not an end in itself.

- Use each of these three major sources of information: the published literature, scientific presentations, and conversations with other scientists.
- To manage the overwhelming volume of available information, prioritize each article (or other source) in relation to your research project, think about its relevance as you read it, and capture your thoughts in written notes.
- Stay up to date with new literature, and ask your advisor to guide you to pertinent past work in your field.
- Allocate regular time slots for reading.
- Choose and use a computer reference database to store and retrieve references to articles and other materials.

References and resources

Dorff, Pat, Edith Fine, and Judith Josephson. 1994. *File . . . Don't Pile! For People Who Write: Handling the Paper Flow in the Workplace or Home Office.* New York: St. Martin's.

Eisenberg, Anne. 1989. *Writing Well for the Technical Professions.* New York: Harper & Row.

Wilson, E. Bright, Jr. 1990. *An Introduction to Scientific Research.* New York: Dover.

17 COMMUNICATING

The importance of communication

Thomas Edison said, "Genius is one per cent inspiration, ninety-nine per cent perspiration." We think that Edison left out an ingredient, and got the proportions wrong: We'd say that success in science is one-third intelligence, one-third hard work, and one-third communication. Anyone who has worked in science for a while knows of people who are smart and work hard but haven't been recognized for their accomplishments because they are reticent, or speak and write poorly about their work. We also know people who give spellbinding talks and come across as smart and imaginative, but when you later think about what they said, there wasn't much there. Flashy PowerPoint slides can—up to a point—mask vacuous content.

Actually, it's arguable that the relation between intelligence and effort, on the one hand, and communication, on the other, is multiplicative rather than additive. That is, if someone is as smart as possible (one-third) and works as hard as possible (one-third) but communicates not at all (zero), their contribution to science may be, not two-thirds of the maximum, but zero. All three factors have to be present at significant levels if the overall contribution is to be significant.

Three key points about communication in science

- You won't really understand your own work until you try to explain it to others. Writing and speaking will clarify and refine your ideas. Talking with others about your work, and writing about it, will get you feedback—critiques of your assumptions, suggestions of better methods, praise for bright ideas and important results—all of which will contribute to your progress and to the quality of the final result.
- Scientific work is wasted unless other people are aware of it and understand it. Research is embedded in a social matrix. You have relied on the work of others as the background for your research, and your work will be unfinished until others in your field know what you have done.

- Professional success depends on being able to communicate professional goals and accomplishments. If you don't communicate your results effectively, not only will they have no impact, but your career will be impaired.

Understand what you're doing, and why

There are various stages and types of communication in research. The most basic, but one that is surprisingly troublesome to many students, is articulating clearly and simply what it is that you're doing, and why it's important. You may need different versions for different audiences—your parents, another graduate student at a party, the faculty members of your thesis committee, and perhaps most importantly, yourself.

Regardless of audience, the fundamental need remains the same: a clearly articulated statement of the aims and significance of your work, and its place in the intellectual scheme of things, emphasizing the forest instead of the trees. Such clarity can be hard to achieve because your own understanding is incomplete and evolving. As a beginning scientist, you may well have been assigned your research project as part of a larger enterprise, and you may not yet have a sense of the whole. But working toward such a sense is one of the most important things you can do.

Expose your ideas to skepticism

Many ethical lapses in science are motivated not by venality but by overzealous attachment to a hypothesis. As we said in chapter 10, one of the most difficult precepts to obey in research is Don't fall in love with your hypothesis. It's hard to come up with an explanation that accounts for your observations, that connects with the broader range of scientific understanding, and that is clever and elegant. It is important to be stubborn in the defense of your ideas, particularly if you are challenging current orthodoxy, but it is just as important to remain properly skeptical, to continue to test your hypothesis rigorously. Achieving the proper balance of stubbornness and skepticism is the toughest challenge in the psychology of research.

Many of the basic precepts of "the scientific method" and "the design of experiments" are directed toward ensuring proper skepticism and rigorous testing of hypotheses. They teach us to recognize that hypotheses will influence the observations we choose to make and the way we assess them. Unconscious bias will be especially great at the limits of observational accuracy, just where a hypothesis might be most stringently tested. If the first experiment comes out the way our hypothesis leads us to expect, we may be tempted not to repeat the experiment or to omit necessary controls.

Apart from the basic tenets of proper experimental design and hypothesis testing, there are two important techniques to avoid self-deception. The first is to keep looking for alternative explanations. If you have two or three hypotheses, then you can concentrate on devising experiments that will distinguish between them, rather than on defending your one idea.

The second technique is to test your ideas with other people—first perhaps with colleagues in your lab, then with a friend in the lab down the hall, then at a research progress seminar in your department, then at a national meeting. It's often much easier for others than for you to see a hole in your argument, to ask whether you've run a control, or to suggest an alternative idea. Even if you successfully defend your position, you will undoubtedly have refined your own understanding of your ideas. The social process of science, grounded in good-natured but challenging criticism of experiments and ideas, is particularly important at this early, informal stage, and can save considerable embarrassment at the formal stage of peer review.

Other reasons to communicate

Not all communications in science are as ambitious or demanding as writing a paper, grant, or fellowship proposal or giving a lecture about your research. You'll also need to communicate less formally, perhaps by e-mail or telephone, to request technical assistance or borrow materials from others, to initiate and maintain collaborations, to persuade reviewers and editors, to exchange information with your advisor and labmates, and to apply for jobs. Each of these situations calls for clear reasoning and expression and appropriate use of language.

Both speaking and writing are necessary

As your research progresses, you will tell people about it in a variety of informal and formal ways. At first you will mainly talk, later you will write. Initially you will discuss your project with your faculty mentor and fellow students and postdocs. After some progress has been made, you will give a presentation to your research group, then perhaps to a departmental student seminar. You may present a brief talk or a poster at a professional meeting. Eventually you will write up the work for your thesis and for publication in professional journals.

Speaking and writing are different modes of communication, with their own requirements and difficulties. But both involve an audience, whose members are usually not as expert, nor as interested, in your research as you are. You need to take special care to rouse and maintain their interest and to make your ideas clear to them. In the next few chapters we will discuss ways to communicate effectively in the various situations that are typical of scientific life.

Take-home messages
- Communication is crucial to your success as a scientist. If no one knows what you've done, your effort is wasted.
- Articulating the importance of your research to different types of audiences will help you refine your own ideas.
- Effective communication will help you learn about what others are doing and get feedback on your own work.

References and resources
Chambers, Harry E. 2001. *Effective Communication Skills for Scientific and Technical Professionals.* New York: Basic Books.

Mogull, Scott A. 2006. *Modern Scientific Communication: The Most Comprehensive Resource for the Development of Scientific Papers, Posters, and Presentations.* 2nd ed. Tificom.

Montgomery, Scott L. 2002. *The Chicago Guide to Communicating Science.* Chicago: University of Chicago Press.

18 GOING TO SCIENTIFIC MEETINGS

The importance of professional meetings

The most likely place for you, as a beginning scientist, to communicate with others outside your lab or department is at a scientific meeting. Even if you don't give a formal presentation, you will have the opportunity to meet a wide range of junior and senior scientists and to discuss research and other professional concerns. This can be a great learning experience. Meetings give you the opportunity to communicate in an intense but relatively non-committal way (in contrast to published work). You can trade ideas, make friends, have friendly discussions and arguments, and give your science a thorough workout in a setting in which everyone is focused on similar goals. In doing so, you help publicize the research being done in your department—one of the reasons departments sometimes provide funding for students and postdocs to attend meetings.

Types of meetings

There are several types of meetings, each with its own characteristics.

- National and international meetings of major scientific and engineering societies can be very large, with thousands or even tens of thousands of attendees. Only the cities with the largest convention and hotel facilities can host these meetings. They are difficult and exhausting to navigate, but "everyone" will be there, and they provide fine opportunities to present your work, learn about recent developments across the entire field, and meet a wide range of junior and senior scientists.
- At the other extreme are small, specialized conferences, often held in attractive locations, with perhaps a hundred people gathered to focus on a particular subarea or problem. These smaller meetings are often the most valuable from a scientific point of view, because of their focus and high level of discussion. They may, however, be limited to scientists who have already established themselves in the field, and thus out of reach for

students and postdocs in training. Nevertheless, if your mentor is invited to speak about your work but can't attend, you might be a fortunate substitute.

- Intermediate in size are regional meetings, which bring together researchers from adjoining states. These are less prestigious and less likely to attract the luminaries of the field, but they can be very useful for gaining experience in presenting and discussing your research. Since they are often held in smaller cities, often within driving distance, they tend to be less expensive, so your research advisor may be more willing to support your attendance.

- Finally, there are local meetings, often on a university or industrial campus. These may be monthly or quarterly meetings of the local chapter of a scientific or engineering society, or they may be ad hoc events arranged by the university to show off their research programs to local industry. They are excellent opportunities to present your research in relatively informal, low-pressure circumstances, and to network with potential employers.

Benefits of attending meetings

Attending a scientific meeting, regardless of the type, affords you certain opportunities:

Get experience making presentations to experts from outside your own lab or department, an important preface to writing up your paper and submitting it for peer-reviewed publication. Get fresh opinions and ideas about your work: Have you tried this? Do you know about X's work? Have you considered this effect? Have you done this control? I tried that, and . . . Such questions and comments are often very valuable.

Build name recognition and your reputation as a promising young scientist. Expand your network of friends and acquaintances, and develop collaborations. Random conversations can sometimes lead to the discovery of fruitful mutual interests. Meet potential employers, or people who work for potential employers (particularly common at local meetings).

National and international meetings, and some of the regional ones, offer significant additional advantages:

- Learn about new developments in your field, often before they're published, about new techniques that may be applicable to your work, and, by attending symposia outside your specialty, about new areas of potential interest.

- Get exposure to the best ideas, the brightest minds, and the highest standards in your field.
- Meet both the leaders in your field, who may be impressed with your research accomplishments and ideas, and up-and-coming researchers, whose experience and day-to-day concerns may be closer to yours than those of the more senior scientists.
- Ask researchers about details of methods or things that have puzzled you in their published work.
- Meet collaborators and friends from distant places, and others from around the world, who can provide useful information and contacts (and may be potential hosts) for overseas travel.
- Interview for positions at job fairs.
- Meet program officers from granting agencies and learn about funding opportunities.
- Learn about the latest research equipment and see the latest books and journals.

Strategies for successful meetings

Use your time carefully and efficiently at a meeting. Read the program beforehand, decide which presentations you must attend—those that bear closely on your own work—and which you should attend if there are no higher-priority conflicts. If possible, give yourself time to explore areas outside your specialty, keeping an eye out for techniques or ideas that might help your own work.

Remember that the formally scheduled sessions are not the only opportunities to interact with others: meals and chat during coffee breaks are often the most productive. Breakfast and lunch dates may be easier to arrange than dinners. Try to arrange such get-togethers ahead of time if there is someone you particularly want to talk with. Otherwise, a casual "Do you have plans for lunch?" or "How about meeting in the lounge to talk about xyz?" is perfectly appropriate. Of course, just a brief chat in the corridor between sessions can also be useful.

Make a list of people you'd like to meet before you go to the meeting. Send them e-mail expressing your interest in getting together, so that they'll recognize your name when you greet them.

Carry a supply of business cards so that people can readily contact you later.

Visit the exposition area to learn about new equipment and supplies. Ask other people about their experience with unfamiliar equipment.

Take thorough notes, so that you can follow up back at your home institution.

Take a break occasionally, and don't overschedule yourself. Information overload can keep you from absorbing anything.

If the meeting is in a big city, try to get a hotel near the convention center. You'll be more in touch with ongoing events. If the meeting is in several locations, try to stay in a hotel close to where most of your action will be. Be aware of free shuttle-bus transportation between meeting sites.

Ask meeting helpers or local arrangements staff—usually at a booth in the lobby—for advice about restaurants, transportation, and interesting things to do if there's a lull in the meeting schedule.

If you're a grad student, use the meeting to explore potential postdoctoral opportunities. Visit posters and attend talks presented by people from labs you might be interested in, try to meet the professors in charge, and talk with students and postdocs from those labs to get a sense of their scientific direction and what your future colleagues might be like.

If you're looking for a real job now or in the near future, use the meeting as an opportunity to schedule interviews with recruiters, as well as to make informal contacts with company scientists.

Posters and platform talks

When you go to a professional meeting as a grad student or postdoc, it's unlikely that you will be an invited presenter giving a plenary lecture. Instead, you'll probably give a contributed platform talk or poster. We'll discuss the details of presenting effective short talks and posters in the next two chapters. For now, suffice it to say that although they're not as important on your résumé as peer-reviewed papers or invited lectures, such presentations are important evidence that you're producing scientific work that's ready or nearly ready for prime time. Moreover, they give you valuable experience in presenting your work and they enable you to get feedback from an audience of experts, giving you the opportunity to refine your work before submitting it for publication. When you apply for your postdoctoral fellowship or first real job, the presence of presentations at meetings on your curriculum vitae will be an important positive factor.

Platform sessions are attractive because the format is familiar. Typically about a dozen speakers in a three-hour period give brief (10–12 minute) presentations on their work in a session with a theme related to their area of research. Each presentation is followed by a couple of minutes of discussion. The audience may be fairly large, but talks lasts only fifteen minutes or so, so you will quickly be off the hook. But the information exchange is often not very satisfactory; opportunities for questions and discussion are severely constrained, especially if your presentation falls in the middle of a long session. Be sure, under those circumstances, to make contact with interested people, so you can talk at greater length after the session, at a meal, or over coffee or another beverage.

A poster session gives you a much better chance to expose your wares to an interested audience. You will generally be located in a section of related posters, so in addition to scientists wandering the aisles, the other presenters are likely to express their interest and ask questions. If someone is interested and knowledgeable, you'll have lots of opportunity to describe and discuss your results, and to answer and ask questions. Others nearby may be drawn into a lively discussion, making the situation even more productive and stimulating. The downside, of course, is that you may end up a wallflower, with few people stopping to read and discuss your paper. However, if you've got interesting results to present, and if you use the presentation tips given in the next chapter, you should attract an adequate audience. Another downside—the cost of success—is that you're trapped at your poster for an hour or so. This restricts your ability to view others' posters and precludes going to concurrent platform sessions. But this is part of the game, and if a few interesting people stop by to talk, you won't regret the tradeoff.

Interacting and networking at meetings

When you attend a meeting, don't just hang out with your friends. You have to interact, to listen attentively, to ask questions, to strike up conversations with like-minded scientists, and if possible to make contact with some of the senior researchers in your field. Don't be brash, but don't be shy. Many others are just as hesitant to approach strangers as you are. But don't be afraid that people won't want to talk with you. Everyone at the meeting shares your interest in the field, and most will welcome good questions, intelligent discussion, and interesting suggestions. Even well-established scientists are generally pleased to be approached by younger ones—it ratifies their prominence and may be the start of an interesting conversation. You might ask your mentor to introduce you to other more senior people.

Occasionally someone you'd like to meet will rebuff your attempts to make conversation in a way that seems rude or cruel. Don't take it personally; it's a poor reflection on them, not on you. Develop a thick skin and turn your attention to someone else.

Remember that communication is a two-way street: listening as well as talking. You want to be noticed but as a thoughtful and intelligent person, not as someone who monopolizes the conversation. Asking good questions, then listening attentively and perceptively to the answers, is a good way both to make a good impression and to learn something.

There's little real distinction between interacting to discuss science and networking to seek a job. As Jensen (2005) defines it, "Networking is the process of establishing links between people with the intent to promote communication for mutual benefit." You may be interested in meeting people who will help you find a job, but they will be more inclined to be helpful if there is a true mutual interaction, if both they and you gain scientific and personal pleasure in getting to know each other. Professional meetings provide a great opportunity make new acquaintances, some of whom may become lifetime friends. If leads for jobs result, all the better, but the real value is in the mutually beneficial personal connection.

Take-home messages

- Understand the different kinds and purposes of scientific meetings.
- Use scientific meetings to get experience making presentations, to solicit feedback on your work, and to ask questions about others' research.
- Expand your network of friends and acquaintances: seek out key figures in your field, rising stars, possible collaborators, and potential employers.
- Attend symposia outside of your specific area of research, to get ideas you might apply to your project.
- Peruse the latest research equipment and recently published books.

References and resources

Jensen, David G. 2005. "Tooling Up: More than Just a Job-Seeking Skill." http://nextwave.sciencemag.org/cgi/content/full/2005/22/17/1.

Tobin, Daniel. 2002. *Capturing Your Next Conference—Getting Maximum Value from Attending a Conference.* Downloadable e-book. http://www.amazon.com/Capturing-Conference-Getting-Maximum-Attending/dp/B00006FC61/.

19 POSTER PRESENTATIONS

Posters have become the most common way of presenting contributed papers at scientific meetings. Meetings have become larger and larger, with more and more people wanting to present their work, while the number of days attendees are willing to spend at them has remained constant (or perhaps even decreased as life has gotten more hectic); as a result, the number of time slots available for short platform presentations has become increasingly inadequate to the demand. At the same time, it has been recognized that three hours of closely packed, fifteen-minute talks is not the best way to convey information or to allow listeners to ask questions of the presenter and share ideas.

Therefore, it has become common to organize massive poster sessions in large ballrooms or exhibit halls, in which hundreds of posters affixed to four-by-eight-foot boards (or smaller) are on simultaneous display for a day—or at least half a day. Presenters are expected to be at their posters for at least an hour or two during that period to answer questions and guide interested viewers through the material. Even in relatively small, specialized meetings, posters have become common; the number of talks during the week will likely be limited to twenty or so, but the remaining hundred or more attendees would also like the opportunity to present their work.

Advantages of posters

The advantages of posters over platform sessions are considerable, and most people feel that they have been a major advance in scientific communication. In addition to providing an overview of some important part of your research, the poster provides an easy opening for discussion. Audience members who are interested can spend as much time they wish discussing and asking questions. You won't be told, "Your fifteen minutes are up. Sorry, we have to move on to the next speaker." Rather, there is time for a real exchange of ideas, for expansion of details, for clarification of obscurities, for suggestions and constructive criticism. Meanwhile, the physical elements of the poster stay put, giving something you and your audience can point to during the discussion.

People who are not interested in the topic are not trapped in their seats waiting for the talk that they really came to hear; they can simply walk down the line until they find other posters that interest them. The atmosphere is informal, with little of the tension attached to a formal talk before an audience.

Posters function a bit like billboards by drawing attention to your work and your expertise, especially if they're attractively designed. They serve this function even if you step away for a moment. They can continue to serve once you're back home, since the poster can often be put up on a corridor wall outside the lab, providing passersby and visitors an opportunity to see what you and your labmates have been doing.

Disadvantages of posters

There are also disadvantages, especially for the presenter. You're trapped at the poster for at least an hour, instead of being done after fifteen minutes. It's polite for the audience in a platform session to stay put for some significant part of the session; here people who are not interested will walk on after, at most, a glance. There may be other posters nearby that you're interested in, but you're not supposed to leave yours during the assigned time—and when you're free to move about, the authors of the other posters may also have left their stations. (Brief excursions, if only to line up time to talk later, are OK.) It's an ego boost if your poster attracts a lot of attention, with many visitors stopping to read and discuss it. You may, however, find yourself trapped for more than the assigned hour, answering the same questions and giving the same guided tour again and again, whereas in a platform session you'd have said your piece once and been free to leave. If, on the other hand, your poster doesn't attract much of an audience, you're left nearly alone for an hour—a real downer. In a platform session you at least have a captive audience, even if they came mainly to hear the other talks.

Preparation of posters

Poster presentations require careful preparation. They have many elements, which must be carefully designed to fit in a very small space. Each element must be clearly intelligible, as must the relation between elements. Good posters accomplish this and are also visually attractive; but they are, in our experience, substantially outnumbered by poorly designed ones.

The most common problem with posters is that they're too hard to read. Presenters can't bear to leave anything out, so they use type that's too small, squeeze sections together, and present a mishmash that's visually unattractive and hard to untangle. The extreme case, which is more common than one would suppose, is to copy the pages of a manuscript and just pin them up sequentially. Remember that most of your audience will be at least ten feet away when they first spot the poster, and only a little closer if they stop to peruse it. It needs to be readily legible at those distances. That's particularly true, of course, of the banner title; but it pertains to the rest of the poster as well. You don't have to present all the details; you'll be there to explain and elaborate. The main job of your poster is to display the major points of your presentation in a visually attractive and communicative way.

Before doing anything else, you should find out the size of the space allotted for your poster. Four feet high by eight feet wide is about the largest you'll find, and three by five feet or smaller is not uncommon. The meeting announcement and call for papers should have the relevant information, or you can contact the meeting organizers. You'll have to accommodate yourself to that size, however inadequate it may seem, or incur the wrath of your neighbors at the poster session if you encroach on their space.

Once you know the size, and have an idea of the various elements (see next section), lay out the elements to scale on graph paper or on a computer drawing program. This will give you an early, and sobering, idea of how little space you have and how wisely you need to use it.

STANDARD ELEMENTS OF POSTERS

Posters have standard elements, which are not much different from a manuscript or slide presentation but which have to be structured in a particular way for maximum effect. There is a standard placement for some of the elements. The banner title goes across the top. It should be legible from at least fifteen feet away (the typical width of an aisle in a poster session), so make the type as big as possible (letters should be at least one inch, or 72 points, high, but larger would be better). Below the title should be the names of the authors and their institutional affiliations, also in large type though perhaps not as large as the title. If your poster has been assigned a number that is keyed to the abstract book, it might be useful to include it in the banner as well. This banner is your advertisement, helping viewers to locate your presentation and to get as clear an idea as possible of what it's about. The same care that you would use in devising a title for a journal article should be applied here—though you might want to stay away from very long titles.

The subsidiary sections of the poster are not so different from those of a manuscript: Abstract, Introduction, Materials and Methods, Results, Discussion, Conclusions, Acknowledgments, and References. The difference is in their placement, the level of detail, and the way they're written.

In a manuscript, the elements proceed linearly, and are read from left to right. On a poster, the top part is less likely than the bottom to be obscured by spectators crowding around the board. Therefore, the most important sections should be at the top, just below the banner. What are those most important sections? We'd pick the abstract, which summarizes the entire poster; the introduction, which sets the stage; and the conclusions, which summarize what you've done. This effectively duplicates the process a reader would follow in skimming an article. The meat of the presentation, the details of methods, results, and discussion, go on the next level, where interested viewers can inspect them when they get closer (the top levels of results and discussion might also go at the top level, if space allows). Normally we expect to read from left to right, but the space constraints of a poster board suggest a vertical arrangement of pages for each section. Each of the sections should have a label in large type—ABSTRACT, INTRODUCTION, and so on—so that viewers can easily find their way around. Pages need not all be the same size. It's helpful if the pages of a given section are tied together by a border or by a common background (perhaps of colored paper), with contrasting space between the sections.

Don't forget to acknowledge your funding sources and the help of anyone who contributed significantly to the project but is not listed as a coauthor.

The references section should be a *selective* bibliography, citing a few key references, including perhaps one or two introductory references that would orient a novice to the field. The URL of a pertinent Web site would also be appropriate.

LEVEL OF DETAIL

A poster should have considerably less detail than a manuscript but somewhat more than a slide presentation. You might think of it as a somewhat overdetailed slide show, with all the slides visible at once and arranged in a comprehensible order. As in a slide show, you wouldn't want to show a very detailed table or graph. Choose just the material that gets your point across. You do know what point you want to make, don't you?

As we will discuss in the next chapter, on speaking, a slide should have relatively little material on it, just enough to summarize the main points;

it's the speaker's job to carry the argument. This would be a good place to start for each poster page; then ask yourself whether more detail is needed at certain points. Remember that you'll be there to explain and elaborate and answer questions; but there will also be times when you won't be there, so the poster will have to stand on its own somewhat more than a slide show.

With regard to how the sections and pages of a poster are written, again it's best to follow the slide show model: Use bullet points. Don't write complete paragraph or long sentences. With figures and tables, put an informative title on top, the illustration in the middle, and the conclusion to be drawn at the bottom of each page. (If the title adequately summarizes the findings, you may be able to omit the conclusion).

Effective communication with posters

These formatting recommendations are driven by the physical and psychological nature of a poster session. Viewers have lots of competing distractions, and they're likely to stay at least a few feet away from your poster. Therefore, you have to present the material in a catchy, well-organized, and legible fashion. The type size, even of the details, should be large enough to be read from a distance—ideally 18 points or larger. There should be enough spacing between lines and between items that they are distinct and the groupings are obvious. The large type size and spacing, along with the limited area of the poster board, will guarantee that you can't put too much material on the poster. Viewers will average just a couple of minutes in front of your poster, so you want to give them the overall message quickly and entice them to stay longer and learn some of the details. Again, you are there to supply those details, so they don't all have to be written down as they would be in a journal article.

A few extra touches can make your presentation even more effective. Consider posting a snapshot of yourself, so people who don't know you can recognize you in a crowd. If you're planning to be at the poster at times other than those on the schedule, put up a note to that effect. Put up a sheet where interested viewers can write their names and addresses to request reprints or preprints. If you have preprints or reprints available and wish to distribute them, bring a stack and put them on a chair next to the poster or in a manila envelope pinned to the bottom of the poster board.

Posters can get pretty fancy these days, now that most researchers have access to page layout programs and color printers. Indeed, strategic use of color can be a very effective communication device. But don't overdo the graphics or use too many colors or fonts. Your university's graphics departments or a commercial print shop can print the entire poster on one sheet, given a page layout or PowerPoint file. This can look very slick, particularly if it's in color, and it allows you to roll up the poster, carry it to the meeting in a mailing tube, and install it quickly with just a few pushpins. This integrated production limits your flexibility to correct mistakes or make last-minute changes, however, and you can certainly do an attractive job without it.

You may sometimes wish to include a video or computer presentation of some dynamic images. Be sure you really need this before going to the extra work of preparing the material and getting the additional equipment. What used to be a "gee whiz" experience is now pretty routine. Would a sequence of stills convey the same information?

GET STARTED EARLY

Give yourself plenty of time before the meeting—at least two weeks, and preferably more—to try out different layouts and levels of detail. Make sure you have adequate supplies: construction paper, adhesives, pushpins, and so on. If you're using complicated artwork or visual aids, especially if you're having a graphics professional prepare them, make sure you give your materials to the artist or graphics lab well before the deadline. Consult friends and other members of your research group (who may be preparing posters at the same time) for advice on what works best.

Proofread each component of your poster carefully—get others to do so as well—and give yourself time to make proper corrections. Last-minute handwritten corrections don't make a good impression.

Presenting your poster

- Never ship you poster in checked luggage; it could get lost. Carry it with you.
- The psychology of presenting a poster is tricky. You and your work are on display, and you may try too hard to draw people in. Or, fearing rejection, you may act too nonchalant and ignore someone who wants to discuss your work. Try to strike a happy medium.
- Stand somewhat to the side of your poster, so that you don't block the view of those who want to see more than the title.
- If you're not talking to someone, look pleasant and encouraging to passersby, but let them start the conversation by asking you a question or making a comment.
- Try not to get so involved when talking to someone that you prevent others from viewing the poster.

- Don't spend too much time talking with your friends at the poster (but don't ignore them, either). Your aim is scientific communication with as many interested people as possible. You can chat with your friends later.
- Be at the poster when you're scheduled to be there and as much additional time as reasonable. If you have a coauthor, arrange to take turns. Or leave a note telling people when you'll be back and some blank notepaper for them to leave a message if they want to reach you.
- If you're comfortable disseminating preliminary data in this way, consider making reduced-format copies of your poster to hand out at the session. This can make it easier for viewers to discuss the work and ask questions rather than spending their time taking notes.

Take-home messages
- Take advantage of the most important feature of poster presentations: prolonged face-to-face communication and critique of your work.
- Design your posters for simplicity and readability.
- Design a logical flow of content.
- Stand by your poster the majority of the time it is on the board to capture as many questions and comments and meet as many people as you can.

References and resources

Block, Steven M. 1996. "Do's and Don'ts of Poster Presentation." *Biophysical Journal* 71 (6): 3527–29. http://www.biophysics.org/education/block.pdf.

Briscoe, Mary Helen. 1996. "Posters." Chap. 9 in *Preparing Scientific Illustrations: A Guide to Better Posters, Presentations, and Publications.* 2nd ed. New York: Springer.

Gosling, Peter J. 1999. *Scientist's Guide to Poster Presentations.* New York: Springer.

20 SPEAKING

The importance of speaking well

The most direct, and most human, way of communicating is by speaking. It's probably the most important way people exchange information, and scientists are no exception. Of course, aside from recordings, spoken thought is evanescent, so science and scholarship put extra emphasis on written communication. But publishing your work is not enough. On a day-to-day basis, we receive and convey information more by speech than by any other modality (probably even including e-mail, though that's getting close).

We might expect, then, that we would all get a lot of training in how to speak effectively in a variety of situations, but that's not what happens. We learn to talk as children, and at various stages of our education we are asked to say something in class. But when it comes to making presentations that means something, where a job or promotion or important project is at stake, many of us—the vast majority, probably—have had no training in how to do it effectively and well, with a reasonable degree of comfort. Those who have gone through Toastmasters testify to its value, but many academics would be ashamed to admit that they had done so, even if they had.

Even more basic is the confidence and ability to comfortably hold a conversation with a stranger. To do this well isn't easy—many scientists lack the skill—but the payoff can be great. Learning to read the individual you are speaking with is critical. Did they understand what you just said? Did it spark their interest or bore them? Are they comfortable with your style of communication? Are you standing too close to them or too far away? Do you express interest in what they are saying by maintaining eye contact and smiling, or do you look away?

These are skills that can be learned and practiced. If you're uncomfortable in such conversations, force yourself to become a bit more extroverted, at least occasionally, so that you can practice speaking with more people. If you're not sure how to do it, observe and analyze people who possess these skills, and, if possible, get one of them to explain the basics to you.

The main message of this chapter is that speaking in a variety of circumstances, formal and informal, is an important part of your professional training. You should learn to prepare carefully and take pride in doing a good job. If you'd like further pointers, we recommend Robert Anholt's book *Dazzle 'em with Style: The Art of Oral Scientific Presentation*, which is full of practical advice.

Ethical issues

Being ethical is just as important in oral presentations as in writing. Always acknowledge your coworkers and collaborators. It's conventional to do this at the end of a presentation, but doing so at the beginning is also appropriate. If someone has done or helped with a particular part of the work, mention them when you discuss it.

You should also acknowledge whoever funded your work. Federal agencies and private foundations alike will appreciate this. If your research was supported by a private company that might have a commercial interest in the results, your audience will feel deceived if you don't acknowledge this support.

If you're using data or a figure from someone else's publication, there should be an acknowledgment on the slide. Likewise if you're quoting from someone's published work.

And of course, be honest in the data and conclusions that you present. Don't pretend that unfinished or preliminary work is more conclusive than it is. Give your audience a sense of the uncertainties, while still presenting a coherent story.

Impromptu talks

Effective speaking means having something to say and making it fit both the occasion and the audience. Having something to say is (or at least seems) obvious—if your speech is vacuous, a knowing audience will soon catch on, and it will be much more damaging than not having been there at all. But just what to say—how much, in what detail and depth—depends on the context. In addition to the formal fifty-minute lecture, there are several other kinds of "talks" that you'll likely have occasion to deliver: the "elevator talk," the "corridor talk," and the "office talk." In each case, someone asks you, "What are you working on?" You want to impress them with your ambition, insight, and accomplishments but have only a limited time.

THE ELEVATOR TALK

Imagine you're in the elevator, riding between the second and sixth floors of your research building, and someone asks what you're up to. You have thirty seconds—maybe fifty words—to summarize your work. Could you boil your answer down to a couple of sentences? What's it about? What's interesting and new? Why do you think it's important? A tough task, but sometime you'll have just a few moments to chat with someone important, and if you come out with something that they can understand and that piques their interest, the follow-up could be significant.

THE CORRIDOR TALK

The corridor talk might be five-minute encounter in the hall with a colleague you haven't seen in a while, who might have a visitor in tow. You have no visual aids or blackboard, just your voice, demeanor, and gestures. You can go into more detail than in the elevator, but not much more. Can you capture the essence of your research, put it in context, indicate the key problems and puzzles, sketch your approach, summarize the results to date and what you hope for next—all in five minutes? You can't go into a lot of detail, but you want your listeners to come away with a good sense of your problem and its scientific importance, of your clever and thoughtful approach to the project, and of the contributions you've made and hope to continue to make.

THE OFFICE TALK

In the office talk, you have twenty minutes and a blackboard but no other props. This is fairly typical for a meeting with a visiting speaker, who goes around the department before his or her talk, learning what the faculty and students in the department are doing. Anyone who has participated in these exchanges, from either side, knows that it's a challenge to stay interested (or even to stay awake, after lunch) while someone drones on about their research. How can you keep the visitor interested, so they will leave with some recollection of what you do, a feeling that you do it well, and a sense that the department and university that has you as a student or post-doc must be a good institution? You have more time, but again, you don't want to get bogged down in details. You want to put the problem in context, then spend some time on what work you and others have done up to this point: what the major unresolved issues are, how you're addressing them, what the progress is so far, and what you think the next steps are. Some well-conceived diagramming on the board may make things clearer.

Just as important, you now have time to engage the visitor in a dialog, to make this a two-way discussion among scientific peers, each of whom has something to contribute. In classroom instruction we call this "active learning," a situation in which the students are not passive bystanders and recipients of information but instead are engaged in trying to formulate, understand, and find solutions to a problem. It's a good way of keeping people interested and getting them involved so they will remember what went on. Of course, you want to maintain some semblance of superior expertise—it's your research project, after all—but discussions with a visiting expert in a related discipline or subspecialty can be both interesting and useful. More than one such conversation has led to ideas for experiments, suggestions that helped to overcome difficulties, or even new collaborations.

In all such encounters, whether thirty seconds or thirty minutes, it's important to quickly gauge the degree of knowledge and expertise of your audience. Adjust the level of your talk to the level of the listener. Your listeners' expertise is the key issue here, not their degree of interest. Your mother is undoubtedly very interested in your work, but unless research is the family business, she'll need a very elementary presentation. A visiting expert may not initially be very interested (he's distracted and just had a heavy lunch), but it is within your power to captivate him with your story.

Formal talks

TALKS TO RESEARCH GROUP

As a grad student or postdoc, your most frequent oral presentations are likely to be to your research group. These occasions can be very valuable, both as practice in speaking and in organizing your thoughts and for the feedback they invite. In this relatively informal setting, you can share preliminary results, try out new ideas before committing them to print, request critical responses, and engage in discussion. This exchange of information benefits the group as a whole; you, as well as your listeners, may be exposed to new ideas.

TALKS TO DEPARTMENT

If you are a graduate student, you will probably be required, once a year or once in your graduate career, to give a talk to your department. The benefits to you as speaker are the same as for a talk to your research group, with the addition that you have the opportunity to make a good impression on faculty and other students.

The audience, too, benefits. Faculty members and students learn what's going on in the department; some information may be relevant to their research, some an addition to their general scientific knowledge. They also value, or should value, the opportunity to ask questions, and engage in discussion. Taking in spoken material provides practice in real-time critical thinking, and some may hope to make a good impression by asking insightful questions. Finally, it never hurts to observe others' presentation practices, good or bad.

THE JOURNAL CLUB

Before you give a talk about your own research to your department or at a national meeting, you will probably give a journal club talk, in which you present and discuss a paper from the scientific literature. This is a special genre, though it has many elements in common with other types of talks. From preparing and presenting a journal club you can learn new and interesting material, which may be useful in your current or future research, by teaching it, and at the same time get practice in the critical evaluation of papers, in presenting material orally, and in answering questions and thinking on your feet.

To prepare for a journal club presentation, you should pick one or more papers dealing with a topic that interests you, is significant enough to interest others, and can be handled in the time available. If you're doubtful about any of these points, check your choice with your faculty advisor or some other trusted source. You should begin preparing soon enough to read the paper(s) several times, check background material when necessary, develop a critical understanding of the paper and its contribution to the topic, prepare adequate visual aids, and rehearse.

Among the common mistakes of journal club presenters, especially neophytes, are choosing papers that are either incorrect or insignificant; failing to understand the articles adequately; and either not properly defending the work against audience members' criticisms or glossing over real flaws.

As you can see, presenting a journal club talk involves developing exactly those critical skills that you will need to properly assess the literature in your chosen field of research.

THE FORMAL LECTURE

The most fearsome type of speaking is the formal seminar or lecture. It may be relatively short—a fifteen-minute platform session at a professional meeting—or it may be an hour-long lecture to your department or a department at another university. Here you have time to plan your presentation, to develop visual aids, and to practice, so there's little excuse for not doing a very good job. However, a very good job is surprisingly hard to do.

As always, the first thing to remember is that your aim is to communicate with a particular audience. You don't want to just stand up there and unload your entire stock of knowledge about your project, you want your audience to understand what you've done and come away with a message they can remember.

THE SHORT PLATFORM TALK

Let's start with the fifteen-minute platform talk. Actually, you should aim for twelve minutes of lecture, leaving the last few minutes for questions. If you've done a good job, you'll get lots of questions from an interested audience. It's frustrating to know that you've sparked interest and a potentially good discussion, only to be cut off because it's time to move on to the next speaker. (An even worse sin in these tightly scheduled sessions is to run into the next speaker's time. You may think that your presentation is the most interesting and important, but most of your audience will not agree, certainly not the speakers who follow who are pushed out of their scheduled slots or denied their own fully allotted time. It's a good way to embarrass yourself and arouse antagonism, which you certainly don't want to do.)

A good thing about a short talk in a platform session is that you can assume your audience is knowledgeable about the topic and has come to pick up the latest information, not to be educated about an unfamiliar area. Nevertheless, you should still devote the first minute or two to orientation, indicating your project's overall importance in relation to the field. Move next to a couple of minutes about your approach, just what you're trying to do, and your methodology. Then about five minutes of results, two minutes of discussion, and a one-minute summary and reprise, and you're done. A fast pace, but you've covered the necessary ground and announced your contribution to the interested world. Hopefully, you'll have time for a more extended talk with the most interested audience members in the hall outside the conference room or over lunch or a beverage at the end of the session.

You should plan for one slide every one to two minutes. For a twelve-minute talk that means no more than a dozen slides. You can flip past your title slide quickly. That should be followed by one that outlines the major issues, then one that outlines what you'll cover in the talk. Then one or two on methods, two to four on results and discussion, one that summarizes the question, approach, and conclusions, and a final one that thanks your coworkers and the funding agencies. That makes eight to eleven slides, which is about right.

THE HOUR-LONG LECTURE

Another important type of communication is the lecture, usually invited, at a conference or at another research institution. This is a significant opportunity to present your ideas, since you will have a claim on the attention of your audience for as much as an hour—more time than most busy readers, except true devotees of your topic, would give to reading one of your papers. You can put your ideas in context, survey a whole body of work—not just one paper's worth—and use all the persuasive devices that go with live, face-to-face communication. If your talk is a plenary session at a large conference, or to a department at another university, you can also expect a broader audience than would generally read your papers. This can help you establish yourself as an important figure in your discipline, not just in your narrow subspecialty.

In a longer talk, you have more time for each stage of the presentation, but you should take care not to run long. If you're told you have an hour, aim for forty-five minutes of formal talk. That gives you a little buffer if you run over and allows for ten minutes or so of questions at the end. Again, if you have made the talk interesting, you'll get lots of questions, from both experts and neophytes. You're probably talking to a more general audience than in a platform session, so you should spend considerably more time on your introduction, setting the scene and explaining your approach. You also will probably be telling a more extensive and complete story, so you will have more results to present. The discussion should also be longer, allowing more opportunity for a discursive analysis of the results and their meaning. How you distribute your time will vary, but you might start by allocating five to ten minutes for your introduction, five minutes for methods, twenty minutes for results, ten or fifteen minutes for discussion, and about five minutes for the conclusion. You should end with a single summary slide, giving the message that you'd like the audience to take away. As with the platform talk, plan on one to two minutes pre slide; thirty slides for the whole talk is a reasonable estimate.

Dealing with questions

During and after your talks, particularly the longer ones, you will undoubt-edly have to field questions from the audience. You should be glad to get them, because they mean that people are interested and have been listen-ing critically. Some, however, show more insight than others. Here's how we suggest you deal with various common types of questions.

Insightful questions or comments. These are the best kind, because they show that someone really understands what you've been talking about. De-pending on the nature of the question or comment, you might say some-thing like "That's an interesting idea. We've looked into it and . . ." or "That's a great question. I'm going to look into it as soon as I get back to the lab. My guess is that we'll find one of two things . . ."

When you don't know the answer. Don't pretend that you do. Say, "That's a good question. I don't have an answer right now, but I'll look into it when I get back to the lab and I'll let you know."

Dumb questions. Sometimes people ask questions that show clearly that they haven't been paying attention or haven't understood the basics of your talk. Try to answer the question in the simplest, clearest way possible. Don't show irritation or try to embarrass them; your audience will react better if you respond kindly.

Redundant questions. If a question is asked that has already been an-swered during the presentation, be patient and address it again, briefly. Perhaps you can quickly go back to the relevant slide.

Questions during the presentation. If the lecture is small and informal, you might explicitly encourage questions during the presentation. Other-wise, it's best to try to hold questions—unless they address some specific point that needs clarification—until the end. If someone asks a question that would be better held until the end or will be addressed later in your talk, say so and ask them to be patient.

Successful speaking

Regardless of the type of talk, there are some common factors that contrib-ute to success. We can group them under three main headings: planning, practice, and presentation.

PLANNING

Outline your talk carefully and thoroughly to develop a clear and logical progression. Clarity is essential. If you don't understand something the first time you read it, you can go back and read it again. When you're listening to a talk, there is only one chance to grasp a point. Make it as easy as possible for your listeners to follow what you want to tell them.

- Begin with an introduction that surveys what you will be talking about and emphasizes the major points you wish to convey. Be sure to state clearly the issue to be discussed. End with a conclusion that recapitulates the main points. As many a teacher of speech (and writing) has said, "Tell them what you're going to tell them, then tell them, then tell them what you've told them."
- Once you've introduced your subject, give some background. Many in the audience will not be experts and will appreciate an orientation to the topic and its context.
- Spare the tedious details of methodology. They can be discussed in the question period or later, in private conversation, if someone is interested.
- Don't just present a succession of data; use intermediate summaries and transitions. Every few minutes, summarize what you've just said and where you're going next.
- Have clear in your own mind, so that you can convey it to your audience, the take-home message of your talk. If listeners remember just one thing, what do you want it to be? What are a few subsidiary points (not more than five) that back up the main message?
- Make clear when you've finished. Don't just stop. Say something like "Thank you for your attention. I'll be happy to answer questions."

PRACTICE

When you have to give a presentation that matters, you should never just stand up and wing it, especially if you're a beginner. Practice! First alone, then with an audience of your peers, friends, and—if possible—your mentor.

Walk through the talk several times. Do the points you want to make come across smoothly? Are there gaps in your argument that become apparent as you speak? If you have a particularly important talk to give (a presentation at a national meeting, or a departmental seminar), ask members of your research group to help you rehearse. Get them to suggest improvements in content, organization, and presentation.

Check your timing. Speak slowly, but don't run over your allotted time, particularly if it will encroach on the next speaker or keep the audience from leaving at the end of a long day. If you realize you are running out of time, skip to the most important slides and summarize your main points.

Don't read your talk or memorize it. Don't read your slides either. Have an outline (your visual aids will probably serve) and use that to guide your talk. You should, however, prepare opening and closing statements, so that the message you present is as clear and cogent as possible.

Allow adequate time for discussion, generally three minutes even for a brief platform talk.

Become familiar with the room before you speak, if you have the chance. Make sure that all the audiovisual equipment you need is available. Preview slides to make sure they're right side up. Check the LCD projector hookup to your laptop. Check for spare bulbs, and give some thought to what you would do if the projector lamp blew out. Check that there's chalk for the blackboard, or markers for the whiteboard.

Don't get flustered if something goes wrong. It's inevitable that once in a while a slide will get jammed or the computer will misbehave. Know your material well enough to improvise while the problem is being fixed. If you need to show a graph, you can probably sketch it adequately on the board, rather than waiting for it to reappear on the screen.

PRESENTATION

Even if your material is carefully organized and you've practiced extensively, you're likely to have a few butterflies when the moment of truth arrives and you're facing your audience. That's OK—it shows you're anxious to do a good job. If you've prepared well, you should be relaxed after the first few moments of your talk. Here are some hints for a successful presentation.

- Don't apologize for being nervous or unprepared or unqualified. Just start talking.
- Humor is OK, but be careful. Don't insult people or groups or embark on a long story that may not be all that funny. If you want make jokes, it's safest to make them at your own expense.
- Academic dress is notoriously casual, but it's better to be a bit too conservatively dressed than to be too informal, particularly if you're young and looking for a job.
- Make contact with the audience. Face the audience, not the board or projection screen. Make eye contact with members of the audience for a few seconds at a time, then move to the next person. (This is

easier to do if you have friends in the audience.) Move around a bit, and use hand gestures. Watch for signs of bewilderment or boredom.

- Don't mumble. Speak clearly, and emphasize the points you want to make by modulating your voice. If your voice is soft or the room is large with poor acoustics, use a microphone.
- When you rehearse in front of your friends or labmates, ask them to point out physical and verbal tics—nervous gestures and repetitive sounds ("like," "umm") of which you may be unaware. Work on eliminating these distracting mannerisms from your delivery.
- Be succinct and to the point when answering questions. Repeat questions if the audience may not have heard them clearly. If a questioner is persistent in disagreeing with you, admit that both of you may have part of the truth and invite a private discussion afterward.
- Be respectful of your audience. Show that you value their time, that you've made the effort to prepare, and that you take the occasion seriously.

Visual aids

No matter what kind of talk you're giving, you're likely to use visual aids: PowerPoint, slides, overhead transparencies, or a blackboard or whiteboard. The first principle, regardless of the medium, is to make sure the text and illustrations you're presenting can be read by everyone in the room, particularly those sitting in the back. Don't crowd too much information into one image; what's important is that it be legible.

For text, use phrases or short sentences. Avoid misspellings, which are distracting and undermine the audience's confidence in your competence. If you're using photocopied graphs and figures, enlarge them so that all of the type is readable at the necessary distance; if you can't, make new ones. Be sure graphs and figures are clearly labeled. Tables should contain only the information discussed in your talk and, again, must be readable. It's often best to make new tables rather than photocopying them from a journal article. If your presentation would be made clearer by an explanatory diagram or flow chart, draw one that is tailored to your argument.

BLACKBOARD OR WHITEBOARD

Sketching on the blackboard or on an overhead transparency is becoming less common, but for an informal talk—or even a formal talk at a small meeting—well-executed hand drawings can be very effective and engaging, especially when it's the general trend rather than the exact numbers that matter. People sometimes get tired of overly slick computer presentations.

OVERHEAD TRANSPARENCIES

Overhead transparencies are handy for less formal talks and for occasions where you must prepare material at the last minute. Here are some pointers for using them successfully:

- Don't cover and progressively reveal information on transparencies. Use overlays instead.
- Stand to one side, so that you don't cast a shadow or block the audience's view.
- Face the audience, not the screen.
- Place a low table or chair next to the projector on which you can stack your transparencies before and after you've projected them.
- Make sure there's a spare bulb.
- Keep your transparencies clean and well-protected in a sturdy box, so you can use them again.

SLIDES OR POWERPOINT

It used to be that less formal talks were given with chalk at the blackboard, or with overhead transparencies, while more formal ones were given with projected slides. Now talks of all kinds are given using computer-based projections. PowerPoint and similar programs allow you to produce images that look finished and professional but, in contrast to slides, can be assembled and modified at the last minute. Since slides and PowerPoint images are similar in terms of their appearance on the screen, we'll discuss them together. A useful resource for effective slide presentations is Michael Alley's book *The Craft of Scientific Presentations.*

Structure of slide presentations. Slides should be readily intelligible and designed to quickly orient the audience. Use one slide for each one to two minutes of your presentation.

- The introductory slide should tell what you're going to do.
- Use background slides to motivate and supply context.
- A diagram that illustrates your apparatus or diagrams the flow of your analytical method may be useful if the apparatus or methodology is not standard. The diagram should be simple, visible from the back

of the room, and not overly detailed. If you've kept the audience interested, the experts who want to know the details can ask questions, and you can elaborate in response.

- Your argument should be carried by graphs, tables, and figures. Visuals are more memorable than words. Data slides should have a title that states in a short, declarative sentence or phrase what the slide is intended to convey. Short texts (no more than two lines each) under each illustration should support the message.
- Use transition slides to summarize the course of the argument. When possible, use pictures or diagrams rather than just words to indicate organization and main points.
- A single conclusion slide should emphasize the take-home message, and ask for questions.

Legibility issues. In slides, clutter is the enemy of clarity. Too much type, too many colors, too much detail in illustrations—each is an impediment to audience members' comprehension. A good rule of thumb is that a slide should contain no more than you could type double-spaced on a three-by-five-inch card: nine lines of forty-five characters is a standard. A figure or table means proportionately less text. Don't try to crowd twenty slides' worth of material onto ten slides. White space will be worth more than an additional fifty words, especially if they're unreadable. Here are a few other pointers:

- PowerPoint's default type is a serif font such as Times New Roman, but a sans serif font such as Helvetica (also known as Arial) is more readable on slides. Use at least 28-point bold type for titles and 24-point type for the body type. Never use smaller than 18-point type.
- Don't use all capital letters except for one- or two-word headings; they are hard to read quickly by shape.
- Be sure that slides are visible under a range of lighting conditions. Black on white and yellow on blue are standard. Many other color combinations don't project well.
- Color can be used to distinguish different data sets. You have to move along briskly, and your listeners won't have time to leisurely consult the figure legend. Are the black triangles ducks and the black squares geese, or vice versa? Give the audience cues that make it as obvious as possible. You might draw a red duck and a gray goose, then color the associated points and lines accordingly. Avoid

using red and green for contrast, since red-green colorblindness is not uncommon. Red and blue are standard.

- Don't use fancy visual effects, animations, or sounds in PowerPoint presentations unless there is a real reason to do so.
- Some presenters use two projectors. This can be useful if you will refer frequently to a key diagram or picture or if you have a series of side-by-side comparisons that won't fit on single slides. But too often this is simply an attempt to squeeze in twice as much information; forcing listeners to zig and zag between projections can also result in sensory overload. Remember that your voice should carry the story—you should be predominantly a story-teller—and the illustrations and words on the screen are there for backup.

Finally, if you are using a laptop computer with an LCD projector, try to set up well in advance and be poised to start your presentation. Test that the projector works with your operating system (Mac or Windows) and that you have any necessary adapter cables. Although the technology is becoming more reliable, it is wise to have a set of backup transparencies.

Take-home messages

- Be prepared to answer the common question "what are you working on?" in formal and informal settings, in one-minute as well as one-hour formats, for scientists in or outside of your field or a lay audience.
- Learn to engage others and to invite and maintain a two-way dialog.
- Plan your formal talks to be informative, clear, and logically organized.
- Practice your formal talks. Ask your advisor and labmates to offer suggestions for improvement.
- Don't memorize your talk.
- Allow time for discussion and practice answering questions, particularly ones you do not know the answer to.
- Take advantage of modern presentation tools such as PowerPoint, but recognize that they tend to distance you from the audience.

References and resources

Alley, Michael. 2005. *The Craft of Scientific Presentations: Critical Steps to Succeed and Critical Errors to Avoid*. Springer-Verlag.

Anholt, R. R. H. 2005. *Dazzle 'em with Style: The Art of Oral Scientific Presentation*. 2nd ed. Academic Press.

Hill, Mark D. "Oral Presentation Advice." http://www.cs.wisc.edu/~markhill/conference-talk.html. (See especially the section by David Patterson, "How to Give a Bad Talk.")

Radel, Jeff. "Effective Presentations." http://www.kumc.edu/SAH/OTEd/jradel/effective.html.

Laskowski, Lenny. "Free Public Speaking Tips." http://www.ljlseminars.com/monthtip.htm.

21

WRITING

Writing is one of the most important things you will do in your scientific career. In the course of your student and professional life, you'll write your dissertation, journal articles, grant and fellowship proposals, and reports of many kinds. Some will be lengthy and complex (the dissertation or a book), while others are short (memos or e-mail), but all are important. In the following chapters we'll discuss writing for particular purposes. In this chapter, we present some general ideas that apply to all kinds of writing: why it's important, how to persuade yourself to sit down and write, and how to write clearly.

Writing about your research

Publishing your research in professional journals and in your thesis or dissertation benefits you directly. It gives you visibility in the profession. It secures your claim to having discovered an important piece of knowledge. It may lead to collaborations and invitations to talk at scientific meetings.

Writing is also an obligation to your research community. Your research is not done until it's been written up, peer-reviewed, and published in the open literature. Only then can it be used and built upon by other scientists.

As you progress through your career, writing is evidence of productivity, justifying continued funding of your research through grants and fellowships and continued provision of space in which to do the work. Success in academic jobs is probably based more on publications than on any other single factor. Promotion, tenure, and higher salary correlate directly with productivity, measured in terms of the volume and influence of your publications. The same will be true if you get a position as a researcher in a government laboratory. Publishing in the open literature may be less important in industry, but you will still need to write good, comprehensive reports on your research progress and plans, in order to justify continued support of your projects.

Productivity—as measured by writing—is important for your future career, for the reputation and funding of your research group, and for science generally. It is through your publications that others will be able to build on the advances you have made. Even at the beginning of your career, you should aim to be prolific but not trivial. If possible, publish at least three papers from your thesis. They will of course be related, but each should be solid and distinct. Your project should be deep enough to allow this. A similar number of good publications should be your goal as a postdoc.

Writing as an aid to thinking

Writing is not just a means of advancing your career. It is also is a way of thinking. It helps to draw out your thoughts, so that you can examine them explicitly. Sometimes things come out that you might not have realized you were thinking about, and sometimes those thoughts are valuable. This is the sort of thing that should go on in your notebook as you plan, and later reflect on, the day's activities. It's productive to have dialogues with yourself through writing, as well as with others through conversations and meetings. Both contribute to making murky things clear, dealing with unanticipated ideas, generating new ideas, and looking at things in new ways. Writing is a way of thinking, and it should be pursued constantly.

Writing thoughtfully but informally in your notebook about results as they're obtained is a wise idea. A colleague of ours says that each experiment should be viewed as a point on a graph or an entry in a table that will appear in a manuscript. That is, data should always be collected and critically examined in light of its final use in a publication. This is made easier if you write about the data as it's obtained: Are there any special circumstances or characteristics that make this result surprising? What new lines of experimentation does it suggest? Have adequate controls been run? Does it fit with theoretical expectations? How might this set of data be fit to an equation? What should be done next? This critical, ongoing evaluation is often neglected, judging by our experience working with students. They tend to gather lots of data in tables before plotting it, and not really to think about what it means until they're ready to write a final manuscript. By then, however, interesting details that could have been noted at the time have been forgotten.

In reporting research there are always a lot of writing chores that will need to be done eventually. They might as well be started at the beginning and completed as one goes along, both to get in the habit of recording information and to get them out of the way. Most obvious is the materials and methods section of a paper. Write a sentence or a paragraph right away about where the reagents came from, how the instruments were calibrated, what data reduction equations and statistics were applied. You should also make an early start on a literature survey. Not all of your survey will be used, but getting started early—not just reading papers but making notes and raising questions about what they say, noting things that seem solidly established and also discrepancies, gaps, and conflicts between results from different labs—is a crucial step in pursuing research. Reviewing, critically reading, and making notes on the literature of your field should be started as soon as possible and continued throughout the course of your research.

A systematic process for writing

NOTES AND ZEROTH DRAFT

Eventually you'll get to the stage where you need to stop researching and start writing. Writing should proceed in three stages: gathering and arranging data and ideas, connecting them in logical order, and refining them for communication to readers. You'll find it easier to write if you keep these stages, which you can think of as zeroth, first, and second drafts, fairly separate.

Consider the work habits of Pulitzer Prize–winning author John McPhee, often considered the dean of nonfiction writers, as described by William Howarth (McPhee 1982). McPhee takes copious notes during his research and interview phase, then sorts the various snippets into bundles corresponding to chapters, puts each chapter's worth into a big manila envelope, and pins the envelopes in order to a wall; this is the zeroth draft. Next he takes down one envelope at a time and writes a first draft of the chapter, marking each envelope/chapter as it's finished, as a visible indicator of his progress through the book. Finally, he revises.

The zeroth draft stage is similar to the "writing as an aid to thinking" described in the previous section, but it is directed specifically toward producing a finished piece of writing. During this phase you'll assemble the notes you made as you reviewed the literature, did experiments and calculations, gathered data into graphs and tables, and wrote descriptions of materials, methods, instrument calibrations and controls, and data analysis. As you do this, you'll come up with more ideas for what the data mean and how the various pieces might be arranged in a logical order. Make notes on these ideas, and group them with the other items into appropriate sections for the piece you're writing, whether chapters of a dissertation or sections in a manuscript or grant proposal.

The noted psychologist B. F. Skinner asserts that if you write enough notes, major papers can be assembled out of material you already have on hand. He also observes that it's not a good idea to write full sentences and paragraphs in the first instance, since it's psychologically harder to rearrange and reassemble them into a finished structure. It may be better to jot down fragments.

FIRST DRAFT

The first draft stage involves arranging the fragments that make up your zeroth draft by topic, and then connecting the topics in sequential order. You'll want to construct an outline to get an idea of how the pieces flow together and to make sure you haven't left anything out. Many of the major word processing programs have outlining views, and there are specialized outliners as well. However, you can do just as well by hand.

The most important structural element of any serious piece of writing is the paragraph. Each paragraph should contain one major idea. So the outline headings below the section level as you begin your first draft should be the basic ideas (topic sentences) to be carried by each paragraph. Under each, list the few points, in sentence form, that will flesh out the main idea of that paragraph.

You'll want to arrange your outline headings, or paragraphs, in an order that will guide the reader to follow your chain of evidence and reasoning. To do that, you'll need suitable transitions between paragraphs, so that the last sentence of paragraph n leads logically to the first sentence of paragraph $n + 1$.

Write the first draft quickly, without worrying too much about refinements or transitions. You can think about transitions later if they don't come out readily. For now, it's most important to get your main ideas and supporting points down in a logical order that carries your argument.

REVISING

Having written a first draft, set the work aside for a few days, then reread and revise. Organizational matters, as well as nuances of phrasing, may become clearer with a bit of perspective.

Consider novelist Stephen King's rule that the first draft should be written with the door closed, the second draft with the door open. That is, the first draft can and probably should be a mess that gets the ideas down but will require substantial editing; at that initial stage, you don't want to be inhibited by the thought of having someone else read it critically. The second draft, however, is written for public consumption; you will initially invite criticism from a few close, trusted readers, and eventually from a broader readership.

Plan to spend at least as much time revising as you did writing the first draft. Your thoughts, and the story you're telling, will evolve as you write. Revise carefully, and more than once. Each rewrite should refine the thought as well as the expression. During revision, try to work on the whole piece at once, especially if it is long, to gain unity of tone and treatment. Try out successive versions on your advisor and colleagues.

Whether you revise on the computer screen or on hard copy is a matter of personal preference. However, changing the medium on which you work may give you a different viewpoint, which can be useful in revising.

Writing discipline and writer's block

Writing demands the same kind of time and energy as doing experiments and calculations. And it may require more self-discipline, since to most scientists it's less exciting. The longer you wait, however, the harder it will be to pull together your thoughts on the completed results and to resist the siren call of the new.

Writing, like most other tasks, is done most effectively if it's done regularly and with some discipline. You expect to come in every day and do an experiment, even if you're tired or not in the mood. Treat writing the same way. Don't put it off until the end of the day, when you're tired, or the end of the project, when you're too rushed to do an adequate job. Set some daily goals for writing—number of words written, time spent, or both—and stick to them.

Consider the British novelist Anthony Trollope, who in the nineteenth century wrote about fifty books while being fully employed by the English post office. (He invented the postal box.) He got up early each morning and wrote twenty-five hundred words at the rate of a thousand words an hour. When you consider that the average scientific manuscript has about five thousand words, plus tables, graphs, and references, you'll see that churning out the words is not intrinsically the rate-limiting step.

The key to productive writing is to keep to a regular schedule. Try to pick a time when you are fresh, rested, and able to concentrate; most people find the morning best. Pick a quiet place where you won't be interrupted (turn off your cell phone and don't look at e-mail), and arrange your writing space so that it suits your idiosyncratic preferences. Plan to spend at least an hour or two there, each day producing a page or more of notes and ideas (zeroth draft), outline and paragraphs (first draft), or finished prose (revisions). Make yourself sit and write, rather than using the time for busywork chores like hunting down references.

Even with a schedule, a quiet place, and the discipline to sit down and write, sometimes the words just won't come. This is called writer's block, and it affects most writers get it at one time or another. Here are some tips on how to get out of it:

- It's often difficult to get started, with a new section of a manuscript or with the day's writing session. So go ahead and start in the middle. Stop today at a moment when you know the first sentence you'll write tomorrow.
- If tomorrow you find that you're stuck, back up a little and review (or even retype) a bit of previous work to get a running start. Or warm up with a bit of writing not related to your main project.
- Write in whatever order the material coalesces. Drafts are meant to be changed. Introductory paragraphs especially often have to be recast or replaced, so don't worry that it's not perfect right away. Just get things down on paper. Prime the pump. Bits and pieces can be rearranged later, either by physically cutting and pasting paper or, less laboriously, on the computer.
- Once you've got started, keep writing. If you're stuck because you're missing a piece of information, leave a blank rather than interrupting the flow. You can look things up later.

Writing clearly

Your research is not complete until it is written up and published, and it will not have the impact it should have unless it written clearly enough that it can be readily understood by its intended audience. Since at this stage of your career you'll generally be writing for an audience of scientists who are reasonably knowledgeable about your area, difficulties in understanding are not as likely to arise from the complexity of your argument as from poor organization and expression of your thoughts.

Writing clearly and effectively is an enormously broad and deep topic, and we can only scratch the surface here. You are urged to consult one or more of the many good books on scientific and technical writing and on writing in general, a few of which are listed at the end of the chapter.

Regardless of whether you're writing a dissertation, a journal article, or a grant proposal, your document will have some predictable elements: titles and section headings, paragraphs, sentences, and words. We'll consider these in turn, pointing out ways to enhance clarity.

TITLES AND SECTION HEADINGS

Your document needs a title, which should identify the field to which the work pertains and the emphasis of your particular research within that field. The challenge is to make the title adequately specific but not overly detailed. The title "Protein Folding," for example, identifies only the field, not your particular contribution. Such a title might be appropriate for a broad review article for a general audience, or a monograph for specialists, but at this stage of your career you are unlikely to be writing either of these works.

If the emphasis of your research has been on application of a new technology to study protein folding, then an appropriate title might be "Use of Laser Light Scattering to Study Protein Folding." If you have been studying the effects of solution additives, then a suitable title might be "Inhibition of Protein Folding by Detergents." If you've focused on the folding of a particular species of protein involved in Alzheimer's disease, then you might get more specific and use "Inhibition of Beta Amyloid Protein Folding by Detergents." The title "Inhibition of Beta Amyloid Protein Folding by Detergents as Studied by Laser Light Scattering" is, however, too long and the extra methodological detail is not the main focus of your work.

Section and subsection headings help the reader navigate your text. Short phrases or single words set off typographically, they are necessary in all but the shortest pieces of text. After reading the title, readers will typically look for the section heading (usually "Summary" or "Abstract") that introduces a brief statement of your work's main points. Beyond the title and summary, each type of writing—dissertation, journal article, fellowship or grant proposal—has its own conventional or appropriate sections. We will describe these in subsequent chapters.

PARAGRAPHS

As we noted earlier, paragraphs are the main structural elements of a piece of writing. Each should carry one major idea that helps to develop the overall thesis of the piece. That major idea is expressed through a topic sentence, which in scientific writing is usually the first sentence of the paragraph.

The other sentences in the paragraph support and develop that major idea. The last sentence, like the first, is in a position of emphasis. It is often used to restate the major idea or to draw a conclusion from it. If possible, it should also point forward to the next paragraph. Making smooth transitions between paragraphs helps the reader to follow your argument.

Paragraphs should be neither too short nor too long. A short paragraph—only one or two sentences—may be used occasionally for emphasis or effect, but generally the idea carried by a paragraph requires several sentences of explanation or elaboration. On the other hand, if a paragraph is very long—beyond, say, three-quarters of a double-spaced manuscript page—then it probably contains more than one major idea and should be broken in two. If readers have trouble identifying your main points, for this or any other reason, they may lose the track of your argument. Three to eight sentences is often considered to be the ideal length of a paragraph, but there is no strict rule; you will have to use your judgment, which will develop as you continue to write.

Don't bury important details in the middle of a paragraph. Consider this paragraph:

> In this paper we report an extensive survey of the solution conditions that affect the thermodynamics and kinetics of the reaction. The important variables include temperature, pH, ionic strength, concentration of ethanol, concentration of sucrose, and type of catalyst. Each of these conditions will be discussed in detail.

The reader is unlikely to remember, or even notice, all the items in the long list of variables. The following paragraph is better, both because the list is shorter (the concentration variables have been combined) and because it comes at the end of the paragraph, where it receives greater stress.

> This paper reports the effect of solution variables on the thermodynamics and kinetics of the reaction. The important variables are temperature, pH, concentrations of solution components, and type of catalyst.

Don't make lists too long; the later items may be overlooked. Put the most important items up front, and try to combine the less important ones in a single item. Bulleted or numbered lists are good for displaying discrete items, but they take a lot of vertical space and may not be allowed or appropriate in a journal article or grant proposal where space is limited.

SENTENCES

Paragraphs are composed of sentences, and just as a paragraph must have a clear structure to convey its meaning, so must a sentence. Sometimes there are definite rules for sentence construction, but at other times you must rely on your judgment to determine the best organization. An article by Gopen and Swan (1990) presents some structural principles that make for clearer flow and easier reading in scientific writing. Their arguments are complex and difficult to summarize, but here are some of their major principles:

- Every "unit of discourse"—a sentence or clause—should make a single point. If you try to make more than one point, you're overloading the syntactic unit, and it may be hard to unravel.
- Every sentence tells a story. Put the subject—the item whose story is being told—at the beginning of the sentence, in the topic position. "Enzymes cleave RNA" says something about enzymes. "RNA is cleaved by enzymes" says something about RNA.
- The subject of a sentence should be followed as soon as possible by its verb.
- The "old" information (the connection with what has gone before) should go at the beginning. This provides a context for the new information that follows.
- The "new" information is best emphasized by placing it at the "point of syntactic closure," at the end of the sentence or syntactic unit.

We urge you to read the full article to see how these principles, when put into practice, can enhance readability. A few other basic suggestions follow.

Sentence length. Long sentences, unless they are constructed very carefully, are usually harder to read than short ones. However, too many short sentences in a row are choppy and awkward. Try to vary the rhythm of your writing.

Giving adequate explanation. Don't baldly assert something that needs explanation or motivation to be adequately understood. Compare "We added compound XY to the reaction mixture" to "Because the reaction would not proceed in the absence of X, we added compound XY, which decomposes under mild conditions to yield X and Y."

Antecedents of pronouns. Pronouns—words like "he," "she," and "it" and "this" and "that"—refer back to an earlier noun or noun phrase. Make sure your readers don't have to guess *which* noun. "The ligand binds near the binding site of the cofactor. *It* can be displaced by raising the salt concentration." What does "it" refer to—the ligand, the binding site, or the cofactor?

Agreement of subject and verb. Subject and verb must be either both singular or both plural.

- The bird migrates (singular). The birds migrate (plural).
- The group of resistors is badly calibrated. (The subject is "group," singular, so the verb, "is," is also singular.)
- Chlorine and iodine are halogens. (Subjects joined by "and" are considered plural, hence the plural verb "are.")
- Either chlorine or iodine is used in the reaction. (Singular subjects joined by "or" or "nor" take a singular verb.)
- The data are amazing. ("Data" is a plural noun; the singular is "datum.")

WORDS

Precision in word use. Work to develop a sensitivity to the precise meanings of words. Do they convey the exact distinction or nuance you intend? Does a word conjure up a proper, or a jarring, image? Compare "Compound X enhanced the growth of the tumor" with "Compound X accelerated the growth of the tumor." It's unlikely that tumor growth would be considered an enhancement; the second version is preferable. Become familiar with, and consult frequently, a good dictionary and a usage guide. Do not rely on a thesaurus alone to find alternative words; the definitions and usage notes in a dictionary will be a better guide.

Redundancy and meaningless phrases. Understanding precisely what words mean will also help you avoid unnecessary repetitions. For example, "retreat" means to move back, so in the sentence "The stain retreated back from the oxidizer," "back" is redundant. Some writers have an unfortunate preference for elaborate phrases, perhaps because they sound more weighty and formal; as you revise your writing, ask yourself, for example, if "at this point in time" couldn't be simplified to "now."

Specificity. Don't settle for "changed" or "affected" when another word or phrase could also indicate the direction or magnitude of the effect. Did the reaction rate "increase" or "decrease"? Was the number of observed mutations "cut in half" or "tripled"? Compare "The drug affected the time of survival" with the more specific "The drug extended the time of survival by an average of six months." Even better, put the magnitude of the effect in context: "The drug extended the time of survival by an average of six months, twice as long as the previously most effective drug."

Elegant variation. Don't feel that you must replace recurring words with synonyms to avoid repetition, especially when doing so could cause confusion. Consider

> The chloride binds near the active site of the enzyme, impeding the binding of phosphate. The inhibitor can be displaced only by raising the salt concentration.

Is the chloride or the phosphate (or something else) the inhibitor? Variation in terminology can create ambiguity.

Active versus passive voice. Should you write "The apparatus was adjusted" or "We adjusted the apparatus"? Sometimes it's a matter of taste. In this example you may prefer to emphasize the object or action rather than the author, so the passive would be more appropriate. In other cases the active is preferable: "A precipitate was observed to form in solution" is wordier and less direct than "A precipitate formed in solution." In more personal writing (memos etc.) or when giving instruction ("Adjust the gain to 10x" rather than "The gain should be adjusted to 10x"), the active is usually preferable.

Nouns and verbs. Many verbs have noun forms, and vice versa. Deciding which to use is sometimes related to the question of active versus passive voice: "We observed changes in solar activity" is preferable to "Observations of changes in solar activity were made," but in combination with a more specific active verb the noun form might be better: "Observations of changes in solar activity revealed . . . " With a less specific verb the noun form may force you to add meaningless words: "We measured . . . " is clearly better than "We made a measurement of . . . "

Noun and adjective strings. Scientific writing often includes long strings of modifiers—adjectives or nouns used as adjectives. A procedure for removing oxidized films from metal surfaces can become "metal surface oxidation film removal procedure"—saving a few words but at major cost in readability. What exactly is being removed in that example? In some cases, connecting the modifiers with a hyphen makes the meaning clearer ("man eating shark" versus "man-eating shark"; "patient record management" versus "patient-record management"). Whether you use a hyphen depends on whether the adjective precedes or follows the noun: "time-dependent reaction" but "The reaction was time dependent." (The hyphen may be omitted, even before the noun, in cases where a misreading is unlikely or implausible; e.g., "high school teacher.")

Negatives. Be sensitive to what words mean not just individually but in combination, especially negative words (not, nor), words with prefixes that reverse their valences (anti-, un-, dis-, counter-), and verbs that refer to diminution or contrary action (reduce, inhibit, counteract). References to, for example, "an inhibitor of antibacterial action" (which may be something that enhances bacterial action) have significant potential to confuse readers.

Jargon. Every scientific field has its specialized vocabulary, and you should not be afraid to use yours. But be aware of your audience. If they're not specialists, you should define the terms at their first use, or use more familiar synonyms. And even (especially) if you're writing for specialists, be sure that you are using the jargon correctly.

Definite and indefinite articles. Nonnative English speakers often have trouble with definite and indefinite articles. "A" and "an" are indefinite articles; the nouns they modify refer to any member of a class. "The" is a definite article, pointing to a particular member of the class. Compare "An engine is a machine with moving parts that converts power into motion" with "The engine produces 250 horsepower." "The" may also modify a unique noun: "the theory of relativity."

Either an indefinite or a definite article can indicate that a noun is generic, referring to a whole class, although in somewhat different senses: compare "An electron has a negative charge" (any member of the class of electron) and "The electron has a negative charge" (all members of the class). A generic meaning can also be expressed with a plural noun and no article: "Electrons have a negative charge."

"That" versus "which." "That" introduces a restrictive clause, one that is necessary to distinguish the word it modifies from other things of its type: "The circuit *that* Professor Schmitt invented is called the Schmitt Trigger." "Which" introduces an unrestrictive clause, conventionally set off by commas, which is not essential to establishing the specificity of the modified word: "The apparatus, *which* we constructed out of old tin cans and string, did not work satisfactorily."

Tricky word pairs. Some pairs of words are often confused; be sure you are using the right one.

- affect/effect: As a verb, "affect" means to influence or produce a change in, while "effect" means to cause something to come into being or to bring something about. As a noun, "affect" is used mainly in psychology to mean a feeling or emotion, while "effect" means a result brought about by a cause or agent.

- comprise/compose: "Comprise" means "to be made up of"; "compose" means "to make up": "The solution comprises water, protein, and salt." "The solution is composed of water, protein, and salt."
- continual/continuous: "continual" means "frequently recurring" while "continuous" means "without interruption."
- further/farther: These can be used interchangeably, but "farther" is generally restricted to situations in which the concept of distance is involved, while "further" can also mean "additional," "more extended," or "more."
- its/it's: "Its" is possessive: "The atom lost one of its outer electrons when bombarded by x-rays." "It's" is a contraction of "it is": "It's ionized."
- principal/principle: "Principal" means "main or most important." A "principle" is a rule or basis for conduct.
- complementary/complimentary: "Complementary" means completing or forming a complement to something. "Complimentary" means praising or approving.
- valuable/invaluable: "Valuable" means worth a great deal. "Invaluable" means extremely useful or indispensable.

Take-home messages
- Writing can help to draw out your thoughts so that you can examine them explicitly.
- Establish a writing schedule and stick to it. Write every day, preferably at a time when you are energetic and rested.
- Use your notebook to document research methods, experimental design, and results; to elaborate immediately on what they mean and how they relate to your overall project and other published work; and to record your thoughts about future research.
- When writing a manuscript, start with a zeroth draft in which you simply list your data and ideas. In the first written draft, work to connect your thoughts in a logical order.
- Don't get stuck on minor details in early drafts. Keep your flow going and add details later in revised drafts.

- Use sections and paragraphs to focus the reader's attention on related sets of information.
- When revising, try to work on the whole piece at once to gain uniformity.
- Pay attention to correct grammar and usage. Consult reference books or more experienced writers if needed.

References and resources

Alley, Michael. 1996. *The Craft of Scientific Writing*. 3rd ed. New York: Springer.

Fowler, H. W. 1996. *The New Fowler's Modern English Usage*. 3rd ed. Ed. R. W. Burchfield. New York: Oxford University Press.

Goldbort, Robert. 2006. *Writing for Science*. New Haven: Yale University Press.

Gopen, George D., and Judith A. Swan. 1990. "The Science of Scientific Writing." *American Scientist* 78 (6): 550–58.

Matthews, Janice R., John M. Bowen, and Robert W. Matthews. 2001. *Successful Scientific Writing: A Step-by-step Guide for the Biological and Medical Sciences*. 2nd ed. New York: Cambridge University Press.

McPhee, John. 1982. *The John McPhee Reader*. Ed. William L. Howarth. New York: Farrar, Straus and Giroux.

Strunk, William, Jr., and E. B. White. 1999. *The Elements of Style*. 4th ed. Boston: Allyn & Bacon/Longman.

Young, Matt. 2002. *The Technical Writer's Handbook: Writing with Style and Clarity*. Mill Valley, CA: University Science Books.

TABLES AND GRAPHICS

Scientific communications, whether they are papers, dissertations, slide presentations, or posters, are rarely all words. The data and conclusions from a research project are almost invariably also presented, in part, in graphic form: as tables, graphs, photographs, and drawings. These components of a communication can present information more clearly, economically, and emphatically than text.

It is often said that a picture is worth a thousand words. What isn't so often said is that a good picture (or table) may be as much work to produce as a thousand words. In this chapter we'll discuss when it's appropriate to use graphics and tables and how to construct them so that they communicate effectively in the formats appropriate for various types of presentations.

Should you use a table or graphic?

You first need to decide what you hope to accomplish in presenting a given set of information. Do you want to provide precise numerical data, to show trends, to emphasize comparisons and contrasts, to provide documentary evidence, to explain complicated structures, networks, or interactions, or to display dynamical processes? Only then can you decide whether the information will be presented most effectively as text, a table, a graph, a photo, a drawing or diagram, or an online video. If you can say what you want to say in a couple of sentences, then you don't need a table or graph; either would occupy more space and take more effort to construct than it was worth. Conversely, if you present data in a graph or table (never present the same data in both a table and a graph), you won't need to repeat the numerical values in the text, but the text should comment on the meaning of that information. If the information isn't discussed, then there's no point in presenting it.

Tables

Tables present specific data and enable numerical comparison between data points. According to Briscoe (1996, 41), a table should be used

- To summarize research findings.
- To document experimental procedure and results.
- To allow comparison of related data sets.
- To enable the reader to make calculations from experimental data.
- To enable the reader to reproduce the experiment.

Only the first three points are relevant when preparing slides and posters; the audience for those types of presentation will have time to assimilate only the main points. The last two come into play in publications, which include more detail and should the reader to make calculations or reproduce the experiment if desired.

Tables should be able to communicate information without reference to the text. They should therefore have informative titles or captions, clearly specify units, and include definitions and a key to any nonobvious abbreviations used in footnotes or headnotes.

Standards for the construction of tables are fairly uniform across journals; these standards serve as good guidelines for other uses as well:

- Values to be compared should be aligned in columns, not in rows.
- The headings of the columns should be informative and include units unless units are defined in the title.
- The independent variable should be in the first column.
- The number of decimal places shown should reflect the accuracy of your measurements and the numerical magnitude of the presented data. While 0.003, for example, might be appropriate, 1,052.003 probably isn't. Align decimal points in columns.
- Columns should not be separated by vertical lines.

These standards are illustrated in table 1, all elements of which are fictitious.

Be selective in the material you include in tables. Think of the table's purpose and how you will use the tabulated data in the discussion and conclusions. If there seems to be a reason for presenting tabulated masses of data, perhaps an appendix or online supplement is the better way to do so.

Don't try to use a complicated, expansive table from a publication on a slide or poster. For those formats, choose and present only the most pertinent data, just what is needed to get your message across. Round off numbers to make it easier for your audience to take in the information quickly; eliminate decimals if possible, and list 103,532 as 103,500 or even (if the values in the column were all many thousands) as 104, with the column

TABLE I. Percentage of animals showing symptoms of infection as a function of time after exposure to influenza strain GH1. All animals had been vaccinated with anti-GH1 (100 mg/kg body weight of 1% w/v anti-GHA in standard saline solution) 10 days before exposure

	BIRDS			RODENTS		
Days	Chickens	Ducks	Geese	Mice	Rats	Hamsters
0	0.0	0.0	0.0	0.0	0.0	0.0
1	4.0	3.0	3.5	5.0	6.0	7.0
2	6.0	6.0	5.5	10.0	9.0	11.0
3	10.0	9.0	11.0	15.0	14.0	16.0
4	12.0	12.0	13.0	30.0	28.0	31.0
5	15.0	16.0	16.0	40.0	35.0	39.0
6	17.0	18.0	19.0	50.0	45.0	47.0
7	18.0	20.0	19.0	55.0	52.0	56.0
8	16.0	19.0	17.0	60.0	58.0	58.0

heading indicating multiples of 1,000. In any case, be consistent. Also consider whether a graph would be more effective than an oversimplified table.

Graphs

Graphs are good for showing changes, trends, and relationships. If the same information can be presented equally well for your purpose in a graph or a table, choose a graph. It is more readily understood and more likely to be remembered.

Graphs for journal articles or dissertations, which have separate captions and allow prolonged examination, should be different than those for slides or posters, where quick apprehension of integrated information is required.

TYPES OF GRAPHS

- Scatter graphs are probably the most common type of graph in science. They are used to display the dependence of dependent variables (vertical axis) on independent variable (horizontal axis).
- Line graphs look like scatter graphs and are commonly used with an independent variable that increases uniformly, such as time.

- Bar graphs are good for making comparisons of numerical values. If you are comparing two or more samples, bar graphs usually show comparisons better, while line graphs show trends with time better.
- Pie charts, which show parts as percentages of the whole, are rarely used in scientific work.
- Various specialized scientific and statistical graphs, such as polar graphs and box plots. See Tufte (2001) and Wainer (2004) for lucid and entertaining examples.

In the old days, graphs were often prepared by professional illustrators, whom many departments or institutes had on staff. Now you are much more likely to prepare your own graphs using one or more of the many software packages available. Don't get carried away with the embellishments these packages offer. Make your graphs as simple as possible to carry the information. For example, don't make bar charts with three-dimensional bars or shadows; the information you're conveying is intrinsically two-dimensional. Be careful with shading, if you use it, since light and dark grey might be indistinguishable when printed or copied.

POINTERS FOR PREPARING GRAPHS THAT COMMUNICATE INFORMATION EFFECTIVELY

Axes. Choose the axis values so that the boundary enclosing a scatter or line graph is approximately square, and line slopes are neither very steep nor very flat.

Don't extend the axes much beyond the extreme values of the variables. It is usually best to start an axis at zero, if the actual value of the variable is of interest, or if two variables with significantly different magnitudes are being compared. If necessary (and if your software package allows) break the axis between zero and a meaningful minimum value. If only relative changes are of interest, starting the axis at a value different from zero is acceptable. (This can be highly misleading when it's the absolute value that's of greater interest. Think, for example, of the stock market graphs published in the newspaper that plot a variation in the Dow Jones Industrial Average from 9,995 to 10,005 to look as large as a major oscillation from 9,500 to 10,500.)

If one or both variables have a range of several orders of magnitude, consider using logarithmic scales. But recognize that small variations—which may or may not be significant—will be suppressed.

If possible, orient the vertical axis label text to read left to right rather than bottom to top. This may not be possible in journals, where columns are narrow and space is at a premium; but it should be done for slides and posters so that viewers don't have to twist their heads to read the axis label. Axis labels should specify units.

Too many ticks clutter a graph. Be sparing, especially in the use of minor ticks. Because ticks represent gridlines, they should be oriented to face into the graph. You shouldn't need more than four or five numerical values and major ticks on each axis.

Axis lines should be a bit thinner than the curves within the graph, so that the curves stand out.

Curves and data points. When plotting curves, use solid, dashed, and dotted lines in order of decreasing emphasis. Avoid using multiple dash styles, which may be hard to distinguish.

Symbols (sometimes called markers) should be readily distinguishable and consistently used within the paper or presentation. Open and filled symbols are most readily distinguished. Circles and triangles are easier to distinguish than circles and squares. Make symbols of an appropriate size for the figure.

Labels and captions. Make the figure caption informative, so the graph can be understood independent of the text.

In a graph for a slide or poster, you will probably want to put a legend defining the symbols inside the graph and a title above it. These elements should be left out of a graph for publication in a journal or dissertation, since they will appear in the caption. In a poster presentation, a title that describes trends in the data is more informative than one that just has a general description of the topic: "Vitamin D Enhances Calcium Absorption" is a better title than "Effects of Vitamin D on Calcium Absorption."

Labels for the curves, if they are included within the graph rather than in the caption, should be large enough to be legible when the graph is reduced to publication size, but not so large as to be disproportionately dominant. The font size should be no larger, and perhaps a bit smaller, than the axis lettering and numbering. If you use labels within a graph, put them near the curves they identify.

Journal guidelines. Graphs intended for publication in journals have some additional constraints. Each journal will have instructions that you should read and follow. Here are some common directions.

Size the graph to fit in one column (or two, if absolutely necessary) of the journal page, without needing to be resized by the printer. A typical column width is 3.25 inches. Symbols and lettering should be distinguishable and legible—neither too big nor too small—at that size. Ten-point Helvetica (Arial) is standard.

If the figures are to be submitted as electronic files, the journal should specify the file format (EPS, TIF, etc.) that they require.

Print the captions (figure legends) on a separate page from the graph, unless the journal's instructions to authors say differently.

Avoid color in graphs for journals. It's expensive to print and shouldn't be necessary if you choose symbols and line styles carefully. For posters and slides, on the other hand, color is usually desirable.

In bar graphs for journals, avoid fine stippling and solid gray tones because they may not reproduce well as halftone images. Use all white, all black, hatched, or crosshatched fills.

Figure 1, which displays the data from table 1, is a graph in need of major improvement before being submitted for use in a journal article. Among the defects of this graph are

- Too many variables (six) being plotted
- Too broad for one column, too narrow for two
- Axis lettering and numbering that is too small to be legible if the graph is reduced to one column
- Axes that don't start at zero and that extend well beyond the data
- Internal gridlines
- Ticks that face outward
- Too many minor ticks
- Axis labels that do not specify units of measurement
- Title and grouped legend within the graph rather than in the caption
- Symbols that are too small and too similar to distinguish readily, especially if the graph is reduced
- Axes heavier than the plotted lines; line weights that would not reproduce well in halftone printing
- Several hard-to-distinguish styles of dashed lines

Figure 2 is an improved version of figure 1. Note that we have grouped the birds and rodents into single curves, with error bars showing the variation within each group. The title and caption for table 1 would serve equally well

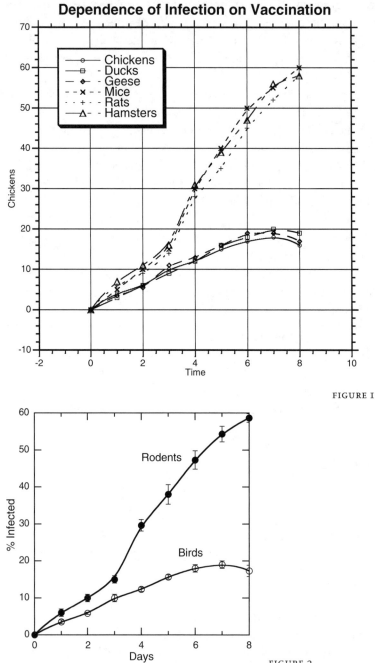

FIGURE I

FIGURE 2

for figure 2, if the author decided to display the data as a figure rather than as a table.

Photographs

Photos may be used for documenting experimental results or for demonstrating their validity. While "seeing is believing" is no longer a given in these days of digital image manipulation, the presumption of honesty means that a picture of a specimen or gel electrophoresis pattern, a photomicrograph of the contents of a cell, or an X-ray of a metal fracture pattern shows what actually was observed.

Photos should generally include a scale marker, so that enlargements or reductions maintain the proper scale information. If possible, submit the photo at the size (one- or two-column) it will be printed in the journal.

If you are preparing a multiphoto composite, be sure that it uses the available space (journal page, slide, or poster) economically. Arrange the components so that the caption is sensibly related to them.

Labels, arrows, and other imposed symbols should be sized appropriately for the image, not so large as to be the dominant feature or so small as to be overlooked.

Manipulations of photographs must not change the information. For example, if the contrast in a photomicrograph of a tissue section is changed to enhance visibility, it should be changed uniformly rather than just in a particular area of interest. To quote from the *Journal of Cell Biology*'s online instructions to authors, "Adjustments of brightness, contrast, or color balance are acceptable if they are applied to the whole image and as long as they do not obscure, eliminate, or misrepresent any information present in the original, including backgrounds. Without any background information, it is not possible to see exactly how much of the original gel is actually shown. Nonlinear adjustments (e.g., changes to gamma settings) must be disclosed in the figure legend."

Halftone photographs submitted digitally must generally have a resolution of at least 300 dots per inch (dpi) at publication size in order to reproduce satisfactorily. Files used for PowerPoint or similar presentation programs typically have a resolution of only 72 dpi, adequate for the screen but not for print. Journals that print a lot of photographs or photomicrographs, such as the *Journal of Cell Biology*, provide detailed instructions that you must follow (see http://www.jcb.org/misc/ifora.shtml#Digital_images).

Drawings and diagrams

Drawings can depict molecular models, cutaway views of mechanical or anatomical structures, or composites of elements that could otherwise be found only in numerous photos. They can be used to emphasize information within an image or to simplify complex structures by removing nonessential or distracting elements.

Diagrams can be used to illustrate complex equipment layouts or complex biochemical pathways, processes, networks, and mechanisms.

If you're good at drawing or can use a computer drawing program, you may be able to produce your own drawings or diagrams. Otherwise, there are skilled scientific illustrators who can help. Be prepared to work closely with one to be sure your meaning is adequately expressed.

Drawings of chemical structures have a highly formalized set of standards, and specialized software exists to help meet them. See the American Chemical Society publications Web site, http://pubs.acs.org/, for instructions.

If drawings are to be used in print, avoid continuous tones, fine stippling, and fine lines. Gradients may get fused and fine details lost in the halftone printing process.

Journals are increasingly accepting color photos and drawings, as the higher information content accommodated by color is recognized and printing costs go down. Color may be rendered either as cyan, magenta, yellow, and black (CMYK), a subtractive process, or as red, green, and blue (RGB), an additive process. Check with the journal about which format they prefer, and how to convert from one to the other if necessary. See Rossner and O'Donnell (2003) for a useful discussion. A striking color photo may be used on the cover of the journal—a nice extra bit of notice for your paper.

Video and animations

Many phenomena of interest to scientists are dynamic, and the Internet now gives us the possibility of putting videos and computer animations of dynamic processes online. This material, if it is part of a scientific paper, should be discussed in the text of the paper and described in an appendix listing supplementary material. Consult the journal to ascertain whether it accepts such animations and, if so, the format in which they should be submitted. Usually there will be restrictions on frame and file sizes. Even if you don't submit animated files for publication, you can still include them in an electronic dissertation and post them on your Web site or the site of your research advisor.

Take-home messages

- Carefully weigh the added value of a table or graph against the effort required to create it.
- Construct informative tables that do not require reference to other parts of the text.
- Choose a graph over a table if the information can be presented equally well in either format.

References and resources

Briscoe, Mary Helen. 1996. *Preparing Scientific Illustrations: A Guide to Better Posters, Presentations, and Publications.* 2nd ed. New York: Springer.

Davis, Martha. 2004. *Scientific Papers and Presentations.* 2nd ed. San Diego: Academic Press.

Miller, Jane E. 2004. *The Chicago Guide to Writing about Numbers.* Chicago: University of Chicago Press.

Miller, Jane E. 2005. *The Chicago Guide to Writing about Multivariate Analysis.* Chicago: University of Chicago Press.

Rossner, Mike, and Rob O'Donnell. 2003. "The *JCB* Will Let Your Data Shine in RGB." *Journal of Cell Biology* 164 (1): 11–13. http://www.jcb.org/cgi/content/full/jcb.200312069.

Tufte, Edward F. 2001. *The Visual Display of Quantitative Information.* 2nd ed. Cheshire, CT: Graphics Press.

Wainer, Howard. 2004. *Graphic Discovery: A Trout in the Milk and Other Visual Adventures.* Princeton, NJ: Princeton University Press.

WRITING AND DEFENDING YOUR DISSERTATION

The dissertation is the capstone of your graduate career. It is both the written embodiment of your contribution to the advancement of your field and a certification of your ability to carry out and write up a significant piece of research. It is the passport to the next stage of your career.

Most books or articles on writing a dissertation take you through the entire postclassroom graduate school process, from thinking about what your research project might be to doing a literature survey, carrying out research and analyzing the results, and writing and revising the dissertation. The earlier steps have been discussed in previous chapters. Here, after one brief detour, we'll focus entirely on the last stage—the writing.

Preliminary oral exam as preparation for the dissertation

Before beginning our discussion of the dissertation, we should comment on its relation to what is known in many institutions as the preliminary oral or proposition. (Some institutions may not have a preliminary oral, or may include the proposition in a comprehensive exam.) Sometimes your proposition will be a direct precursor to your dissertation research. In other programs, the rules may say that this proposal can be some variant of your proposed dissertation. In yet other programs, you will have to propose something completely different. An important aspect of the preliminary proposal is that it gives you practice in formulating a research question that is interesting to the field, neither too ambitious nor trivial, and feasible using available tools.

The preliminary oral proposition should contain three parts: a statement of the problem, background information with literature survey, and description of methodology. If the proposition corresponds closely to your eventual dissertation, these should transfer directly into its first three chapters. If the rules of your graduate program require that the proposal be related to but different from the dissertation project, you may still be able to reuse the literature survey and the materials and methods sections.

What is a dissertation?

According to the American Heritage dictionary (fourth edition), a thesis is "a proposition that is maintained by argument." A PhD dissertation is a formal document that presents the evidence and arguments in support of a thesis. Sometimes a dissertation is called a thesis, and in fact the dictionary treats the words as nearly synonymous.

A dissertation should present original work and be a substantial contribution to the field. It should present not just facts or data, but also critical analysis of findings—whether your own or findings drawn from the literature. The thesis should follow logically from the data and their analysis.

The dissertation should be written in a formal style and a consistent vocabulary, with each statement expressed clearly and unambiguously. The vocabulary should be standard in the field, with any unusual terms defined at their first use and (if there are a lot of them) in a glossary.

Lovitts (2005) carried out a survey of faculty in several fields who had extensive experience in supervising PhD dissertations and serving on dissertation examination committees. These faculty agreed that most dissertations fell in the "very good" category; they rated a very few "outstanding" and a small number "adequate" or "inadequate." According to this survey, a very good dissertation

- Is solid
- Is well written and organized
- Has some original ideas, insights, and observations, but is less original, significant, ambitious, interesting, and exciting than the outstanding category
- Has a good question or problem that tends to be small and traditional
- Is the next step in a research program (good normal science)
- Shows understanding and mastery of the subject matter
- Has a strong, comprehensive, and coherent argument
- Includes well-executed research
- Demonstrates technical competence
- Uses appropriate (standard) theory, methods, and techniques. (Lovitts 2005, table 1)

These are the characteristics you should aim for, though nobody would fault you if you were more ambitious.

Standard components of a dissertation

The scientific dissertation, a formal document with a long history, has evolved certain standard components: introduction, literature review, theory, methods, results or analysis, and discussion or conclusion. The article by Lovitts provides a useful summary of the attributes that are desirable each component (2005, table 2). If you organize and write your dissertation bearing these characteristics in mind, you should find your path to a very good dissertation considerably eased.

COMPONENT 1: INTRODUCTION

- Includes a problem statement
- Makes clear the research question to be addressed
- Describes the motivation for the study
- Describes the context in which the question arises
- Summarizes the dissertation's findings
- Discusses the importance of the findings
- Provides a roadmap for readers

COMPONENT 2: LITERATURE REVIEW

- Is comprehensive and up to date
- Shows a command of the literature
- Contextualizes the problem
- Includes a discussion of the literature that is selective, synthetic, analytical, and thematic

COMPONENT 3: THEORY

- Theory applied or developed is appropriate in its level of detail, logically interpreted, well understood, and aligned with the question at hand.
- Author shows comprehension of the theory's strengths and limitations.

COMPONENT 4: METHODS

- Methods applied or developed are adequate to the problem, described in detail, and aligned with the question addressed and the theory used.

- Author demonstrates an understanding of the methods' advantages, disadvantages, and limitations and shows how to use the methods.

COMPONENT 5: RESULTS OR ANALYSIS

- Analysis is appropriate, aligns with the question and hypotheses raised, shows sophistication, and is iterative.
- Amount and quality of data or information is sufficient, well presented, and intelligently interpreted.
- Author cogently expresses the insights gained from the study and the study's limitations.

COMPONENT 6: DISCUSSION OR CONCLUSION

- Summarizes the findings
- Provides perspective
- Refers back to the introduction
- Ties everything together
- Discusses the study's strengths and weaknesses
- Discusses implications and applications for the discipline
- Discusses future directions for research

Formatting the (almost) finished version

Once you've written and revised your dissertation to the point where it's ready to be reviewed and defended, you'll print it out in the format required by your graduate school and distribute printed copies to your committee members. Further revisions will probably still be needed, so it's not quite time to have your magnum opus bound for posterity. But you should give your committee members a version that—in the unlikely event that they found nothing to revise—could be sent to the bindery upon inclusion of the signature page.

Increasingly, graduate schools allow submission of the dissertation in electronic form. Formatting rules still apply, though they may be different from paper rules. An electronic dissertation is converted to a PDF file and submitted to ProQuest for deposit in its repository.

Check with your graduate school about format requirements, whether for paper or electronic submissions. The instructions may prescribe such details as type of paper, margins, page numbering, type size, line spacing (usually double), paragraph indents, placement of footnotes, captions, tables, and figures, printer quality, use of correction fluid (usually forbidden), and other seemingly picky things designed to ensure that your dissertation will be an attractive, readable document decades and centuries from now.

Typically, a dissertation has to contain the following elements.

- Title page
- Signature page
- Abstract
- Dedication
- Acknowledgments
- Table of contents
- List of tables
- List of figures
- Text (the components listed in the previous section)
- Bibliography
- Appendixes

You may also be expected to follow a discipline-specific style for bibliographic citations and other details. In other cases, you should consult recently accepted dissertations to see what styles are used in your program. A widely used style guide is Turabian, *A Manual for Writers of Term Papers, Theses, and Dissertations.*

Copyright and intellectual property

If your university is in the United States, copyright for your dissertation will be registered with the Library of Congress. Copyright is granted by the government to regulate the use of a particular form in which an idea or information is expressed: the words you used, the diagrams you draw, and so on. Copyright protection does not cover the actual ideas, concepts, or techniques contained in the copyrighted work. You, as the author of the dissertation, will own the copyright. However, any novel ideas or techniques that you developed, intellectual property that may be worthy of patenting and licensing, will typically be owned by the university in which you did your work. You will be asked to pay a fee for registering the copyright for the dissertation. You are not required to do so, and your work will be copyrighted in any case. However, it is difficult to challenge its use by others legally unless you have filed, because not having filed may be taken as lack of intent to defend your copyright.

If the university decides that your work has commercial potential and is worth patenting, it will own the patent but you and others involved in the work (typically your advisor, perhaps other researchers in the lab group) will be the inventors. Patents by themselves are not worth anything. Their value derives from patent holders' power to control the use of an invention. They could commercialize and market products based on the invention themselves, but more commonly the invention is licensed to a company or companies that will take it to the stage of commercialization. Income from licenses is typically shared among the inventors, their department, and the university, often on an equal thirds basis.

Including your published articles in your dissertation

In science and engineering, it is common—and desirable—to publish your work in journal articles or conference proceedings while your research is still ongoing, before you're ready to write a complete dissertation. This is different from standard practice in other disciplines, where the dissertation is written first and then turned into a book or a series of articles.

A question then arises: If the work has already been written up, peer reviewed, and published (or accepted for publication), why go to the effort of rewriting it for the dissertation? Why not just include the articles verbatim? This is indeed a reasonable question, and many graduate schools permit inclusion of published work by the PhD candidate as part of the dissertation. Check the rules of your graduate school and program regarding formatting and acknowledgment of previously published work. Two issues will certainly need to be dealt with.

First, if you are going to include your published journal articles, or even smaller but substantial pieces of published material (such as figures or tables), you need to get permission from the copyright holder: the publisher of the journal in which each article was published. The journal's front pages or Web site should have the necessary contact information. Increasingly, copyright transfer agreements (in which you transfer your copyright to the publisher) explicitly allow for this sort of use by the author, in which case you may not have to ask for permission.

Second, you were probably not the only author of your published articles. The author list likely included your research advisor and perhaps other collaborators. The dissertation, however, is supposed to be your own work, a testament to your ability to carry out a major piece of research on your own. How can you reconcile this conflict? Common sense should be the guide. If the work reported in an article was mainly yours—in conception, execution, analysis, and writing—then it belongs in your dissertation even if others contributed and were listed as coauthors. If, on the other hand, you were listed as an author because you made some contribution to a collaborative project but were not the major contributor, then the article should not go verbatim into your dissertation, though your contributions to it might well be woven into the dissertation in other ways.

Even if you do include published articles in your dissertation, you will want to be sure that the dissertation is more than the sum of its parts. You should include, for example, a more thorough and discursive literature survey, one that provides an integrated overview of the whole topic, not just the parts that were covered in your articles. You may also expand the sections on materials and methods, including information beyond what fit in the limited space of the journal article but that could be helpful for you or others wishing to extend your work. Most important, be sure that your introduction puts your entire research project in context and that your conclusion integrates the findings from all the work you've done.

Getting it done

If the dissertation is the capstone of the graduate experience and the passport to a professional career, why do so many PhD candidates find writing it to be so hard? There are several reasons. First, it's a big job, probably the biggest writing project most students have done or will ever do. Second, it comes after four or more years of graduate work, when burnout is an increasing possibility; it's easy at this point to get discouraged and lose motivation, self-discipline, or time-management skills. For a third, other demands of life—family, a new job offered or pending, finances—are likely to be pressing. Despite these obstacles, writing and defending the dissertation in a reasonable amount of time is entirely feasible. Tens of thousands of graduate students do it each year. In this section, we discuss some useful time-management and motivational stratagems. Refer back to chapters 14 and 15 for a more thorough discussion of these techniques.

SCHEDULE YOUR WRITING TIME

Writing your dissertation needs to be approached systematically, with a feasible schedule rigorously adhered to. It's most sensible to work backward from the time you'd like to graduate, whether to move on to a postdoctoral position or a regular job or to accommodate some transition in your family life. You also should make note of deadlines for submitting your dissertation, scheduling examining committees, and so on. Some universities have deadlines once a semester; others allow submission at any time. Some require that the dissertation be approved before the commencement ceremony can be attended; others allow attendance at commencement before all requirements are met.

With these dates in mind, set yourself a deadline for completing your dissertation and its defense that's challenging but realistic. Take into account the time your committee members will need to read your draft (at least a couple of weeks) and assemble for the final exam. Note that conference and vacation schedules may make it particularly difficult to bring them together during the summer. Allow time for revising based on suggestions made at the exam, to produce the final version of the dissertation. And allow a few days for the mechanics of rounding up signatures, having copies made, and getting final graduate school approval.

Then figure out how much you'll have to write each day in order to meet that deadline. For example, if the typical dissertation in your program is 150 double-spaced pages (including tables, figures, references, and appendixes), you need to have your dissertation ready in reasonable draft form for review by your committee in four months, and your advisor wants a month lead time, then you have ninety days to produce a decent draft. This translates to a little less than two pages a day, but bear in mind that you'll need to revise your first draft before giving it to your advisor, so you might aim for about four pages a day—say, a thousand words or the equivalent. That's entirely feasible.

Your task will be made easier if you have, as we've recommended, written up major pieces of your research as you went along, for journal articles or presentations at meetings. With that preparation, the process of planning, and executing research, analyzing the data, and writing it up, which students in other disciplines may have to handle in bulk, will have been broken into smaller chunks for you.

WORK WITH YOUR ADVISOR

Remember that your research advisor is a key figure in your progress toward the dissertation. The quality of your dissertation will have almost as much impact on his or her reputation as on your own. As you develop a deadline and a schedule for completion, check with your advisor to be sure that he or she thinks it's realistic, and that your research is at a stage where you're ready to write it up and move on. There are often disagreements about this, so be prepared to negotiate.

Once you've agreed with your advisor about the timetable, discuss how much they would like to see in preliminary draft form. Ideally, your advisor will be an active participant at each stage of the writing, looking at a detailed outline, a draft of each chapter, and then the full draft dissertation. But some advisors are busier than others or prefer to give their students more independence in writing up to the final stage. Just be sure that you have a clear mutual understanding of how much help and advice you will get at each stage, and how much time your advisor will need to review each draft. Review your writing schedule with these factors in mind.

USE EFFECTIVE WRITING STRATAGEMS

Some of these pointers on how to write effectively were discussed in more general terms in chapter 21.

- Set up desk space where you will be relatively undisturbed and can keep your writing tools, notebooks, and reference materials spread out.
- Use your table of contents as an outline, to find the most logical arrangement of topics.
- Write a rough draft first. Don't worry about getting all the wording right the first time.
- Write the things that are easiest first, such as the materials and methods section.
- Print out drafts. You see things differently on paper than on the computer screen. The increasing thickness of the paper stack will give you positive feedback about your progress. If you are doing several versions, print each draft on paper of a different color. Or at least make sure that the header or footer indicates the draft, so that you don't get confused.
- Reread the material after a break, when it will be easier to critically evaluate what you have written.
- Back up your work on another hard drive (or CD or DVD) and/or your university's central server. Printed drafts also protect against loss of your efforts if your hard drive crashes; as a last resort you could scan the hard copy back into electronic form.

KEEP BALANCE IN YOUR LIFE

Inevitably, some other activities will have to give way during the time that you are writing your dissertation. But try to keep some balance in your life.

You can't write effectively for twelve hours a day, or probably even for eight. Most professional writers write for four or five hours a day and spend the rest of their working time doing necessary mechanical things (photocopying, checking references, etc.). If you set a reasonable daily goal of time spent or pages written and do that writing first thing each day, you will finish in time. After you've done your daily writing, you can do the things needed to keep up with the other priorities in your life.

Give yourself some time to relax. Build some breaks into your schedule—perhaps one day a week when you don't write, or a couple of days off after each chapter—but keep to the schedule. View the breaks as rewards to which you can look forward.

TAKE ADVANTAGE OF PROFESSIONAL HELP IF NEEDED

If you're having serious trouble keeping to your schedule or are experiencing writer's block, consult your university's writing center or counseling service. They have lots of experience helping people just like you. Your problems are not unique, and they're solvable.

Defending your dissertation

Once you've written a draft of your dissertation that you and your advisor agree is ready for defense, you must actually arrange for the defense. This involves choosing the members of the examining committee (which may have been done earlier in your graduate career, depending on your program and graduate school rules), giving them copies of the dissertation in time for them to read it and approve it as ready for defense, and defending the dissertation in a final oral exam.

CHOOSING THE COMMITTEE

You will probably have an opportunity to choose, or at least suggest, members of your examining committee, in consultation with your advisor or the director of graduate studies. Your school may have rules requiring, for example, that one or two committee members be from outside your major field. These will likely be faculty from whom you have taken courses. For the other members, you'll have to consider the balance between subject matter expertise and personal supportiveness. If possible, pick members who embody both virtues; but if you have to choose, make sure you have at least some who are personally compatible. You will want committee members who have previously shown a willingness to take the time to advise you about your project, help you solve problems, and suggest new approaches.

Most faculty members are fair, and even if they have quarrels with your advisor they usually won't try to take it out on you. If it is known that a faculty member picks on students to get at their advisors, try to avoid having them on your committee.

Be sure that your advisor is on your side and comfortable with your results. She or he may or may not be the chair of your committee but should provide moral support and may find ways to help you through difficult situations in the exam.

GIVING THE COMMITTEE THE DISSERTATION DRAFT

You will give your committee the draft of your dissertation a couple of weeks before your defense, and the readers will have to sign off that it is ready for defense. Some will read it carefully, others will not. Don't be offended by the latter; they're busy people. But when you present your thesis, assume that they have read it. Try to avoid getting involved in any predefense disagreements between your advisor and other members of the committee about how the dissertation should be written. It should be a joint product of you and your advisor, with the other members of the committee reacting rather than being in the driver's seat.

THE FINAL ORAL EXAM

If you have the opportunity, attend the public portions of one or more final exams before your own, to see how they run and what kinds of situations the examinee faces. Analyze what might have gone better, so that it will go better when it's your turn.

Ask some of your student and postdoc friends to sit through a dry run of your presentation and to ask the kinds of questions the examining committee might ask. Get suggestions about how you might have presented and answered better, and take them to heart.

Sometimes you give a public presentation of your thesis, after which you and the committee will go into private session for the examination. By that time, you should be warmed up and comfortable. In other graduate programs, the whole process may involve just you and the committee. In that case, try to reserve the first fifteen to thirty minutes of the exam for your presentation of your project; this will give you a chance to overcome initial nervousness and to demonstrate your mastery of the topic. If the exam begins with questions coming from every direction, you may never catch your balance.

You should view your oral exam as an opportunity to discuss your research with people who are basically interested in and well disposed to what you've done. You know more about your project than anyone else, but the others will have questions and insights that may throw new light on your project and results. Be glad, not defensive. View it as an opportunity to have a scientific discussion among peers. If a question is asked that you weren't prepared for and can't answer readily, say something like "That's a very interesting point, which I hadn't considered. I'll be sure to think about it." (Of course, if the question uncovers a serious error, then you'll need to address it before the final version of the dissertation is printed.) Don't be afraid to say you don't know, if you don't. Ask for permission to speculate. Wrong speculation will not give the same negative impression as incorrect statement of facts.

FINISHING UP

Assuming that you pass your final oral exam (nearly everyone does), you will still have a few details to wrap up. Your examining committee may have some suggestions for minor revisions of the dissertation, ranging from correction of typos to clarification or reanalysis of some results. Make those revisions promptly, and have them approved by the committee if required. Print the requisite number of copies of the final approved version of the dissertation (and perhaps a couple extra for personal use). Following graduate school rules, insert the approval signature page and the copyright page in the official copy, and have the copies bound.

Now you're done! Congratulations! You can move on to the next stage of your career and your life.

Take-home messages

- Browse through recent dissertations by students in your program to get a general idea about their structure.
- Plan a feasible timetable for writing your dissertation and stick to it.
- Get feedback from your advisor.
- Try to get as members of your dissertation committee faculty who have shown a willingness to help you in your research project. Consult with your advisor about committee membership.
- Attend the public parts of others' final oral exams.
- Practice a dry run of your final exam with your peers.

References and resources

Bolker, Joan. 1998. *Writing Your Dissertation in Fifteen Minutes a Day: A Guide to Starting, Revising, and Finishing Your Doctoral Thesis.* New York: Henry Holt/Owl Books.

Davis, Gordon B., and Clyde A. Parker. 1997. *Writing the Doctoral Dissertation: A Systematic Approach.* 2nd ed. Hauppauge, NY: Barron's.

Lovitts, Barbara M. 2005. "How to Grade a Dissertation." *Academe,* November–December 2005, 18–23.

Mauch, James E., and Namgi Park. 2003. *Guide to the Successful Thesis and Dissertation: A Handbook for Students and Faculty.* 5th ed. New York: Marcel Dekker.

Sternberg, David. 1981. *How to Complete and Survive a Doctoral Dissertation.* New York: St. Martin's Griffin.

Turabian, Kate L. 2007. *A Manual for Writers of Term Papers, Theses, and Dissertations.* 7th ed. Chicago: University of Chicago Press.

Zerubavel, Eviatar. 1999. *The Clockwork Muse: A Practical Guide to Writing Theses, Dissertations, and Books.* Cambridge, MA: Harvard University Press.

24 WRITING A JOURNAL ARTICLE

A paper in a peer-reviewed scientific journal is the culminating stamp of approval for your research. It signifies that your work has been reviewed by experts in the field and found meritorious: it is correct, original, and advances the field. If the journal is one of the standards in the discipline, it will be read by most practitioners, so your work will become known to those to whom it matters. The journal's contents will be indexed by bibliographic citation services such as Medline, Inspec, and Current Contents, so that scientists looking for developments in the field will be led to your paper. Without at least one journal article to your name, you will have a hard time getting a postdoctoral position or a job based on your research credentials, or competing for fellowships or research grants. The more papers you have published in good journals, the better your chances of achieving a satisfactory career in research.

In some disciplines a paper in a volume of conference proceedings is equivalent to a journal article. The keys criteria are whether the paper is peer-reviewed and whether the volume is indexed by bibliographic citation services.

The scientific journal article has developed a standard structure over several hundred years. Its parts, in conventional order, are title, authors, abstract, introduction, materials and methods, results, discussion, acknowledgments, and references. There may also be appendixes, and usually some figures and tables. To write a paper that adequately presents your work is not a trivial task, but this standard structure can make the job easier by breaking it into manageable pieces. In the following sections we shall consider each in turn.

The classic book on the subject is Robert A. Day's *How to Write and Publish a Scientific Paper*. Although the book is oriented toward the biological sciences, it has value for all scientific fields. Day presents a great deal of useful information, along with many humorous but instructive examples of how *not* to write scientific prose. Another good reference, more oriented to the physical and engineering sciences, is Michael Alley's *The Craft of Scientific Writing*.

Standard parts of a scientific paper

TITLE

The title is, in some respects, the most important part of the paper, because it is the major determinant of whether people will read it. Most scientists scan the tables of contents of the major journals in their field, or read lists of titles delivered to their desktops by online services. Unless the title accurately describes the contents, the paper is likely to be overlooked by those who should read it. Abstracting and indexing services also use keywords from the title to compile and classify articles. If the title is inadequate, the paper may be improperly indexed and lost in the general mass of scientific literature.

An adequate title describes the contents of the paper with specificity and without extra words. Name the specific species, locations, techniques, or other attributes that characterize your study. The title "Structure of Enzymes in Bacteria" is too general. What kind of structure? Which enzymes? Which bacteria? "A Study of Atomic Structure of Ribonuclease in *E. coli*" is considerably more specific, but the useless words "A Study of" should be omitted, as should similar phrases ("Investigations of . . . ," "Observations on . . ."). If the technique is important, include it in the title: "X-ray Structure of *E. coli* Ribonuclease." (Practitioners will recognize that X-ray structure determination is at the atomic level, so that phrase can be omitted.)

Day (1998) points out that improper word order is a common mistake in titles. He gives the example "Mechanism of Suppression of Nontransmissible Pneumonia in Mice Induced by Newcastle Disease Virus," which seems to imply that the mice were induced by the virus. The intended meaning, of course, was that the pneumonia was induced by the virus; a better title would be "Mechanism of Suppression of Nontransmissible Pneumonia Induced in Mice by Newcastle Disease Virus."

Many journals ask for a list of keywords, in addition to those in the title, to aid in indexing. A good keyword list can make it easier for interested readers to find your paper. You will also be asked to provide a short title (usually less than sixty characters) to be used as a running head. Choose one that accurately describes the paper, even if you can't include all the keywords.

AUTHORS

The title page of your manuscript should also include the names and addresses (including e-mail) of the authors. The journal's instructions to authors will specify the required formatting. Since there may be many scientists with the same first and last names, it is desirable for indexing purposes to include a middle initial. The institutional address will also help to distinguish among the many John Smiths who publish in the scientific literature.

Although people change their names for many reasons, it is desirable to keep your name the same, if possible, on all your scientific publications. If you decide to publish as, for example, Victor A. Bloomfield, do not submit other articles as Victor Bloomfield or V. A. Bloomfield. Consistency will help ensure that a computer search of the literature turns up all of your papers.

The question of who should be listed as an author, and in what order, is often a vexed one, discussed at the end of this chapter. However, for purposes of submitting the paper to a journal, one of the authors will have to be designated as the corresponding author, who will handle correspondence about the manuscript, proofread the edited version, order reprints, and authorize payment for reprints and page charges. This person will most commonly be the research advisor, but the matter should be decided before the paper is submitted.

ABSTRACT

If potential readers are sufficiently intrigued by the title of your paper, they'll next turn to the abstract. This is particularly true for journals available online, where the publisher will typically allow free access to the titles and abstracts of articles, but may charge (or require authenticated access through your university library) for the full article. It's therefore in your best interest to write an abstract that describes the content of your paper concisely and accurately, to enable the reader to decide whether the full paper is relevant and worth reading.

At least as important, the abstract will probably be the first thing read by the reviewers to whom the editor sends your manuscript to decide whether it is suitable for publication in the journal. If your abstract does not do a good job of capturing the reviewers' interest, the chances that your manuscript will be recommended for publication are diminished. Abstracts are also used by editors to assign appropriate reviewers. Thus, a poorly written abstract might damage your chance of getting a well-informed review.

The abstract should be a brief summary of the paper, generally 250 words or less, containing the following elements:

- Context: the importance of the question being studied
- Objective and scope of the research
- Overview of the main methods used

- Summary of the results
- Conclusions derived from the study

The abstract must be self-contained, because it is published, along with the title and authors, separate from the full paper by abstracting services. It should not contain references to the literature or to figures or tables, and should avoid unfamiliar abbreviations or acronyms that will be meaningful only when read alongside the main body of the paper.

INTRODUCTION

The introduction is similar to the abstract, though not as brief. It should present the rationale for the study, indicating the context, nature, scope, and importance of the problem. It should summarize the pertinent literature, not just listing papers and authors, but also indicating the major results in such a way as to make clear what's currently unknown or contested. This literature summary should establish the significance of your paper by demonstrating the need for your research. The introduction should then indicate your objectives and the main methods you have used to address them, along with the reasons for using this approach.

The introduction should end by stating your major results and conclusions. A novelist might approach the story differently, but in science writing "giving away the secret at the beginning" helps to orient your readers to the purpose of your paper, to let them know where you are heading. Your aim in writing a paper is to communicate your results and conclusions as clearly as possible. Stating them in the abstract and the introduction, and more fully later in the paper, is the way to do that.

The abstract and introduction, and sometimes the final version of the title, may well be written last, although they come first in the physical manuscript. It's usually easier to write them once you've gone through the mental exercise of developing your thoughts in the rest of the paper.

MATERIALS AND METHODS

A basic tenet of science is that published work must be reproducible by other qualified investigators; this is essential to establishing its validity. The major aim of the materials and methods section of your paper is to provide enough detail that your work could be repeated and verified if necessary. You might test the adequacy of your draft of this section by showing it to colleagues and asking whether they think they could use it to reproduce your work. They may notice a glaring omission that completely escaped your attention.

The materials and methods section is usually broken into subsections, addressing, for example, materials, instrumentation, assay methods, study site (for field studies), and data reduction methods. The sources of purchased equipment, supplies, and material should be given in detail: name, model number, company, city, state. If equipment was modified or constructed specifically for the study, it should be described in enough detail that another competent researcher could duplicate it. Life science and medical journals, in particular, often have particular requirements for describing biological reagents and human or animal subjects. Consult the instructions for authors of the journal to which you intend to submit the paper.

If a material or method is standard, you only need name it and give a literature reference. However, if your sample preparation or other method is unpublished or modified from the standard, you must provide the detail needed to reproduce it.

Describe precisely how measurements were made, and indicate the precision of the measurements. You can use common statistical methods without comment but should provide references for less common ones; don't detail them in your paper, however, unless you have developed them specifically for this work.

Be sure that you do not include results in the materials and methods section. Conversely, do not present additional methods in the results section. Readers should be able to locate things where they expect to find them.

RESULTS

In the introduction you tell why you did the work, in the materials and methods section you tell how you did it, and in the results section you present your findings. The findings should be presented as clearly and concisely as possible.

Unless you have very few data or observations, you should not present every measurement, but rather summarize them with appropriate statistical parameters: mean, standard deviation, number of observations, and perhaps others depending on the nature of the data.

If there are only a few such parameters, they may be presented in the text. Otherwise, one or more tables or graphs will be more appropriate (see chapter 22). In no case should you repeat in the text data that are given in tables or graphs. Use the text to point out significant features and help the reader understand the meaning of the data. In so doing, avoid excessive verbiage. Don't write, "It is clearly evident from fig. 3 that the reaction rate was independent of salt concentration." Instead, write, "The reaction rate was independent of salt concentration (fig. 3)."

DISCUSSION

In the discussion, you should interpret the results, not repeat them. Your task here is to tell what your results mean and to assess their limitations and significance. A good discussion section should

- State your conclusions—principles, relations, and generalizations shown based on the evidence in the results section—and summarize the case for each as clearly as possible.
- Point out exceptions to and lapses in correlation with general trends, and note unsettled points. Be straightforward; if you try to conceal or obscure such difficulties, reviewers and readers will notice.
- Discuss how your findings and interpretations agree or disagree with previously published work.
- Connect your results with the objectives of the study, and point out the significance of your work, including any theoretical implications or practical applications it may have. Leave the reader with a clear understanding of why your work matters.

In a few cases, speculation that goes beyond the conclusions that are directly supported by your results may be warranted. But such speculation should be brief, cautious, and avoided in most cases. Also avoid promising that future work will follow up or extend the results in the current paper, unless the promised work is already done. Things may not work out as you expect.

ACKNOWLEDGMENTS

The acknowledgments section should briefly express gratitude for scientific and financial assistance with the research. If Doctor X provided a special reagent, and Professor Y contributed some useful interpretive ideas in a conversation, it's good manners to write, "We thank Dr. X for providing the serum, and Prof. Y for valuable discussion." It's probably best to check first that X and Y are comfortable with the acknowledgment, especially if it might imply that they endorse the content of the paper. It is also common practice to note external grants, contracts, or fellowships that supported the research or researchers.

REFERENCES

The references section, along with citations in the body of the paper, is where you acknowledge the sources of information in the literature upon which you have drawn. It gives readers the directions they need to locate that information for their own purposes, and it acknowledges that your paper rests on an extensive foundation of previously published work.

Each scientific journal has its own slightly different conventions for listing and citing references; you'll need to consult its instructions for authors or a recent copy of the journal for the details. Each entry in the reference list should include the authors' names, year of publication, title of the article (some journals do not list the title), name of the journal (generally abbreviated), and volume and page numbers. For books, the publisher and city of publication should also be given. In some journals, the reference list is arranged alphabetically by last name of the first author; in others, numerically in order of citation in the manuscript. Citations within the body of the manuscript typically consist of either the authors' names and date, in parentheses—for example, "(Bloomfield and El-Fakahany 2008)"—or numbers corresponding to those in the reference list, enclosed by parentheses or brackets, or superscripted.

If you use (as you should) bibliographic database software, it will have the proper formats for the major journals, and you can make modifications if your journal isn't included. This can save you much time and effort in formatting the original manuscript, changing the citations if you need to submit it to a different journal, and using the references for other purposes such as a dissertation or another paper. Remember, though, that such databases require careful data entry; if any elements are not in the correct fields, the documentation generated may be useless. We repeat, you should always reread your entire manuscript, including all references, before submitting it.

If you include verbatim quotations of phrases, sentences, or paragraphs from the literature, you must not only give the citation (with page number) but also enclose the material in quotation marks.

APPENDIXES

Appendixes contain information that is too detailed for the main body of the paper, but which may be of interest to a few specialists (e.g., complicated derivations or extensive data compilations). Some journals publish the appendixes online but not in print.

FIGURES AND TABLES

Most scientific papers will contain figures and tables. Proper preparation of these items is discussed in chapter 22.

Present or past tense?

It's often difficult to decide in what tense to write a paper. In general, things that are accepted to be facts or established knowledge are written in the present tense: Such and such is true. Things that you did, which are not yet part of the published literature and are therefore not certified as facts, are written in the past tense: We did such and such. Thus, most of the introduction, which describes the state of current knowledge, should be written in the present tense. The experimental heart of the paper—abstract, materials and methods, and results—should mainly be written in the past tense. The discussion, however, describes what you now think is true, so it should be mostly in the present tense. This issue is lucidly described by Day (1998, 207–9).

Choosing a journal

As the first paragraph of this chapter implies, not all journals are equal. Some are considerably more prestigious than others. At the top of the list are *Science* and *Nature*, two journals that try to report the most innovative research from a wide range of scientific disciplines. A first-authored paper in *Science* or *Nature* is generally considered the gold standard in scientific publication; it will be of great benefit to your career.

Each discipline has its standard journals, which almost all the practitioners read or at least scan. Journals are often rated by their "citation impact factors": the number of times that an average paper in the journal is cited by other papers. A publication in a journal with a high impact factor is generally more highly rated by search committees than one in a journal with a low impact factor. The other side of the coin is that it's usually easier to get work published in a journal of the latter type. You should always aim to publish your work in the most highly regarded journals. It's often worth submitting first to such a journal even if the odds of the paper's being rejected are high (space is at a premium in these journals, and they can't publish everything that's submitted even if it's reasonably good science). You may get some useful comments from the reviewers, and you can always submit the rejected paper to a lower-ranked journal.

Be careful of two things, however. First, if the highly ranked journal tends to take a long time to review, and ultimately reject, your manuscript, your submission to a lesser journal may no longer be timely—you may even be scooped. Check on the journal's reputation for promptness or delay before submitting by consulting colleagues who have experience with the journal. Second, be sure that the lower-ranked journal is indexed by the major citation services. New journals are being started all the time, and it may take several years for them to develop a record of publication that meets the standards of the indexing services. If your paper is accepted and published in such a journal, you'll have an entry in your bibliography—which is better than nothing—but your paper is likely to languish in obscurity and have little or no impact on the field.

As you can see, judgments about where to publish are tricky. You should aim your best work at the best journals, but more routine, follow-up work may usefully be published in lower-ranked journals. Your advisor should be knowledgeable about such matters, so you should benefit from their experience in this regard. Of course, all coauthors should agree about where to submit an article. If there are disagreements, the advisor should have the final say (or, if the advisor is not a coauthor, the lead author should decide).

Formatting and electronic submission

A journal's instructions to authors will generally give detailed directions for formatting the manuscript. Some common standards are that the manuscript should be printed double-spaced, with one-inch margins all around, to enable interlined comments by the reviewers and the copy editor. Words should not be hyphenated at the right margin, because the hyphens may appear within a line when the manuscript is typeset.

Many journals now require that manuscripts be submitted electronically. This speeds the review process considerably, reduces or eliminates mailing costs, and allows editing and typesetting of accepted manuscripts using the electronic file, reducing time and typos introduced during rekeying. The formatting requirements for electronic submission are generally the same as for paper, but the journal may require that all components of the manuscript (figures and tables as well as text) by submitted as a single PDF file for ease in sending to reviewers.

Working with coauthors

WRITING AS A TEAM

Most science these days is a team effort, so your paper will likely have one or more coauthors. At the very least, your advisor will probably be on the author list. We'll assume that the project is largely your effort, so you will be given the responsibility for writing the first draft of the manuscript and coordinating the editing and assembly of successive drafts.

Research advisors have a wide variety of preferences about how to receive drafts of manuscripts from their students. Very few will want to be given the raw data and to write everything from scratch themselves. The great majority will expect the student to write the first draft (or perhaps an outline first) and will then work with the student on successive revisions. This is in part because the advisor is busy, but more importantly because practice in writing is a crucial part of students' education.

If there are other coauthors, ask them to contribute their sections (e.g., writing about their particular technical methods or theoretical analysis, or supplying graphs or photos) to the first draft. Incorporate these contributions with your own writing and assemble a reasonably presentable first draft to give to your advisor and coauthors. (It might in fact be a second draft, if your first draft has large chunks of free writing and just getting things down, as we recommended in chapter 21.) Proofread it, run it through a spell-checker, and, if English is not your native language, consider asking someone more fluent in English to help with textual revision. Or ask a coauthor or a labmate to act as a preliminary editor. This depends to some extent on how clean a draft the advisor wants to work with.

When submitting subsequent drafts, indicate clearly where changes have been made (by lines in the margins or different type style), so your advisor and other coauthors can see readily what they need to be particularly attentive to. These drafts may be made and distributed either as computer files or on paper, depending on lab preference.

Once the manuscript is reasonably complete and ready for submission, give your advisor and each other coauthor a printed copy of the manuscript, glossy photos or computer prints, and computer files of the manuscript, the bibliographic database, graphs and the tables from which they were composed, and computer drawings. All of you are responsible for the paper as coauthors, and all should have copies of the final product.

ISSUES IN COAUTHORSHIP

There are numerous ethical issues that arise when you are writing a paper. Some, such as fraud or plagiarism, arise whether you're a sole author or working with coauthors. Others, less damaging to the fabric of science but annoying and potentially damaging to the careers of young scientists, are unfortunately common when you're writing with coauthors. Here are three of the main questions, with our answers to each. These answers represent commonly held standards of scientific ethics, but (as with many ethical issues) application of standards to particular cases is often difficult.

Who should be listed as authors? Only those who made active and substantial contributions to the planning, execution, and analysis of the work should be included as authors. As Day (1998, 23) writes: "Colleagues or supervisors should neither ask to have their names on manuscripts nor allow their names to be put on manuscripts with which they themselves have not been intimately involved." Those who made lesser contributions might be listed in the acknowledgments.

In what order should authors be listed? The most defensible scheme is listing in order of scientific (not financial or institutional) contribution to the work, but different fields have different customs; some list the authors alphabetically. Perhaps the most common arrangement is to list the authors in order of contribution, except for the lab director, whose name appears last. This arrangement has the effect (which may or may not be justified) of making the first and last authors the most important. In any case, the authors and their order should be agreed on before the research begins, and changed only for good reason and with unanimous agreement.

How much should each listed author know about the contributions of the other authors? In principle, all of the authors should be able to explain and defend any aspect of a paper to which they have affixed their names. In practice, when the work has involved several collaborators contributing a range of disciplinary expertise, the principle becomes somewhat unrealistic. It does seem fair to expect that the collaborators will have spent some time educating each other on their various areas of expertise as those areas bear on the particular research. At the least, as Day (1998, 24) says, "each author should be held fully responsible for his or her choice of colleagues."

A thorough, detailed discussion of ethical issues in scientific publishing, emphasizing those arising from coauthorship, is Marcel LaFollette's *Stealing into Print.*

Take-home messages
- Divide your research into publishable segments.
- Aim for top journals in your field but be realistic in matching the quality and impact of your work with journal standards.
- Ensure that the title and abstract of your article provide an informative summary of the content of the manuscript.

- Provide comprehensive and fair coverage of the relevant literature.
- Make sure that your conclusions are strongly supported by the data you present and by published work.
- As a rule, write the introduction and discussion in the present tense, and the abstract, methods, and results in the past tense.
- Involve other coauthors in the writing process and get their feedback regularly.
- Pay attention to the ethics of authorship.

References and resources

Alley, Michael. 1996. *The Craft of Scientific Writing.* 3rd ed. New York: Springer.

Day, Robert. 1998. *How to Write and Publish a Scientific Paper.* 5th ed. Phoenix, AZ: Oryx Press.

Katz, Michael J. 2006. *From Research to Manuscript: A Guide to Scientific Writing.* Dordrecht: Springer.

LaFollette, Marcel C. 1992. *Stealing into Print: Fraud, Plagiarism, and Misconduct in Scientific Publishing.* Berkeley: University of California Press.

25 WRITING FELLOWSHIP AND GRANT PROPOSALS

Whether you are a graduate student or a postdoc, you might have to write and submit a proposal for a fellowship to support your research. We advise you to apply for your own fellowship even if you are supported by your advisor's grant. Going through the process will teach you the ropes. Furthermore, receiving a funded fellowship is a distinct honor and an impressive addition to your curriculum vitae. When you finish your postdoctoral training and move on to an independent research job, you will undoubtedly have to start writing proposals in one form or another, whether you do research in an academic institution, industry, or government. The most common type, which you will need to write and write frequently, is a proposal to fund your research program. This could be a grant proposal to be submitted to a funding agency or a research plan to be considered by an internal review body. Regardless of the format and purpose, successful proposals have characteristics in common.

Do not let being a rookie discourage you. The fact is, you have some distinct advantages over veteran investigators: You are youthful, enthusiastic, and full of energy; you likely have more exciting and cutting-edge research ideas; and you are probably better trained in modern research methodology. Most important, reviewers generally sympathize with junior investigators and will be inclined to support you if you put forward your best idea and present it in a clear and convincing manner.

However, you should expect that putting your first few proposals together will take you longer than it takes a veteran. So start early, and plan for numerous revisions. Schedule your writing so that you can provide a draft to colleagues for criticism a couple of weeks ahead of the proposal submission deadline, and remember that you institution's sponsored project office will probably need a few days or a week to process and submit the application.

Identify potential sources of funding

When you consider applying for an individual fellowship to finance your graduate or postdoctoral research, the most likely sources are federal agencies such as the National Institutes of Health (NIH) and the National Science

Foundation (NSF). Depending on your research area, you might also consider funding from the U.S. Department of Agriculture, Department of Energy, or Department of Defense. There are also various local and national foundations and associations that support graduate and postdoctoral research, for example the American Heart Foundation, March of Dimes, and the American Cancer Society.

GrantsNet provides an extensive directory of funding agencies, both federal and private (http://www.grantsnet.org/search/fund_dir.cfm). Other online search tools that can help you identify possible sources of funding include the Sponsored Programs Information Network (http://www.infoed.org/new_spin/spin.asp; access requires an institutional subscription), Community of Science (http://www.cos.com; again, an institutional subscription is required), and the Foundation Center (http://fdncenter.org; free of charge). These search engines will provide you with all the information you need about the funding agencies identified in your search: eligibility criteria, deadlines, maximal amount and duration of support, and guidelines for writing the proposal. It is OK to send the same proposal simultaneously to different funding agencies, with the understanding, of course, that you will accept funding from only one.

Understand your audience

A realistic perception of how fellowship and grant proposals are reviewed is of the utmost importance for being in the right frame of mind when you write. Novices often assume that each proposal is reviewed by one or more experts in the specific area of the proposed research. They imagine that reviewers are highly compensated for their efforts; committed to understanding what exactly the applicant is proposing, even if it requires rereading the proposal; and predisposed to give the applicant the benefit of the doubt, despite missing information, obscure and ambiguous prose, or poorly stated goals and benefits (because *all* research is worth doing).

Most of these assumptions are incorrect. Without this realization, as painful as it might be, you will have only a slim chance of hitting the bull's eye. Reviewers are as busy as you are with their own work. They too must divide their time between a multitude of daily tasks: teaching, exams, research, committee meetings, and, not to be overlooked, writing their own research proposals.

Another false assumption is that grant review and funding is mainly a test of one's good fortune, that if researchers simply keep applying eventually they'll get a hit. One cannot categorically deny that luck plays a role in the process. There are many factors you can't control: who is assigned to review your proposal, their mood when they conduct the review, who attends the particular round of reviews when your proposal is being considered. You must, however, work to minimize the role of luck in determining the funding decision. Your best bet is to propose a novel and important research question, and to present it so clearly that it has instant appeal to the reviewers. As Thomas Jefferson once said, "I'm a great believer in luck, and I find the harder I work the more of it I have."

Some applicants submit rushed and imperfect proposals just to see where it will fall on the funding scale, hoping, if nothing else, that the reviewers' critiques will help them write a better version next time around. This is certainly a grave mistake. Do not think that reviewers will eventually fund your revised proposal just to avoid seeing it again in every round of reviews. This never happens. Instead, your first poorly written application may negatively color their reaction to other proposals you submit in the future. You obviously do not want this sort of reputation. Moreover, some funding agencies limit the number of revisions of a given proposal that may be submitted. The NIH, for example, allows only two revisions.

Understand the funding process

It is important to have a general understanding of how proposals are processed and how funding decisions are made. While these steps take place after you submit your proposal to the funding agency, realizing what is involved is of immense help in writing successful grant and fellowship proposals.

When proposals are received at a funding agency (assuming they arrive by the specified deadline), they are usually checked for completeness and compliance with agency instructions. Large federal agencies like the NIH have dozens of review committees organized according to areas of research. Program administrators assign each proposal to the appropriate review committee, which in turn assigns it to the members of the review panel best qualified to review it. More than one reviewer is usually assigned to each proposal, and one is selected to serve as the primary evaluator. This person is usually, but not always quite knowledgeable in the specific topic of the application; some review panels are made up of only a handful of individuals who are responsible for reviewing proposals across a very wide spectrum of research areas. Even when the primary reviewer is an expert, it is easy to imagine that the secondary and tertiary reviewers might have little or no knowledge of the specific topic of your proposal. In other cases all members of the review panel may be asked to read and evaluate each proposal, regardless of their particular backgrounds and expertise.

Proposals are mailed to the assigned reviewers, who are usually given two to four weeks to complete their evaluations and provide numerical scores. All members of the review panel then meet to discuss the entire set of proposals under consideration. Sometimes this task entails handling more than a hundred proposals in one or two days, allowing only a few minutes to discuss each proposal. Certain funding agencies (e.g., the NIH) exclude applications that fall in the bottom half of the numerical ratings from discussion in order to focus on those with greater potential for funding. During the review of each proposal all assigned reviewers are asked to declare their scores or degree of enthusiasm. The primary reviewer then summarizes the proposed work and the strengths and weaknesses of the proposal, and the secondary and tertiary reviewers indicate whether they agree or disagree with the primary reviewer's assessment. Following a short discussion, the reviewers give their final scores. Other members on the panel score proposals on the basis of this discussion. Applications are ranked according to average scores, and this ranking is used in allocating the available money. In most cases you will receive a summary of the reviewers' critiques of your proposal, along with the average score, independent of the funding decision. This process can take up to six months (and sometimes longer) from the time you mailed your proposal.

When you bear this sequence of events in mind, you will recognize the importance of markedly impressing the reviewers in the shortest possible time. This can be accomplished only by presenting your clever research idea in the clearest way possible. This clarity must be readily apparent to the expert and the uninitiated alike.

Understand the criteria used for proposal evaluation

Most funding agencies provide their reviewers with guidelines for evaluating proposals. These lists share a number of common themes; in essence, reviewers are usually asked to answer these five fundamental questions:

What research questions are proposed? Why were these specific questions chosen? Are the questions significant and innovative? How much knowledge stands to be gained if the proposed research is done?

How will the research questions be answered, and to what depth? How carefully has the applicant considered alternate tests to validate results? Have all necessary controls been thought out?

Who will be doing the research, and who will be supervising them? Does the applicant demonstrate appropriate training, knowledge, and past productivity. If you are applying for a predoctoral research fellowship, having published one or two abstracts of your results at national meetings would suffice, while several peer-reviewed papers would be expected in support of a postdoctoral fellowship. Whether you are a pre- or postdoc, the productivity of your advisor and his or her track record with past trainees matters a lot. Quality of training is reflected in job placement of past trainees.

Where will the research be done, and does the institution have adequate resources? Will the institution where the project will be conducted provide adequate technical resources and intellectual support. Does it have the core facilities and modern equipment required for your research? What is the quality of its library? Is there a critical mass of investigators who work in your areas?

So what? What is the potential benefit? How does the proposed project relate to the mission of the funding agency. All such agencies have to justify their annual budgets by providing an account of how the research they funded has impacted their specific goals. Some funding agencies, for example the National Science Foundation, specifically ask for justification in terms of the research's broader social impact.

Have a positive attitude

The most important tool in grant writing is an optimistic attitude. There is a lot of grant money out there. Outstanding proposals successfully compete with all others in the heap and get funded. Spend significant time contemplating which of your ideas would be best to use for the research proposal. You should be sincerely excited about this particular idea. Your enthusiasm will show between the lines of the proposal and will in turn fuel reviewers' enthusiasm for what you are planning to do. Again, put yourself in the reviewers' shoes before you start describing your idea in writing. Prepare yourself psychologically to write for very busy reviewers, only a minority of whom are experts in your research field, who have little time to spare. Be prepared to write many drafts of the proposal, not stopping until you produce a flawless final version that you can be proud of. This naturally necessitates starting early.

Understand the elements of successful proposals

Two things, more than anything else, will determine your chances of success: (1) the research idea you are presenting and (2) the clarity with which it is presented. It is not possible for a research proposal to succeed with only one of these components. The jury is still out on the question of which is more influential. Poor writing will almost always damage a brilliant idea, but even the cleverest presentation cannot turn a bad idea into a good one. To be on the safe side, ensure that your proposal possesses both essential qualities.

COMING UP WITH A NOVEL, COMPELLING, AND INFORMED IDEA USING THE "SCIENTIFIC METHOD"

The scientific method need not apply only to scientific projects. The term more generally implies using a structured way of thinking that results in identifying focused, novel, and worthwhile research questions.

Start by considering all of the background literature that addresses the research direction you are planning to pursue. Identify specific gaps in existing knowledge, then prioritize them according to their significance for the field. Focus on the most important question or set of questions and contemplate how providing answers would significantly advance knowledge. If you are convinced that there is evidence, from the background literature or preliminary experiments, to support a likely answer to the question, then this question could be posed as a specific hypothesis. The process of deciding on the best idea to pursue will usually require careful and methodical thinking, which over many days or weeks will result in gradual refinement and focusing of your research idea.

Never propose to pursue a research question simply because it has not been addressed before. This is by far the poorest justification that could be provided to reviewers. In the absence of a clear, educated, and systematic justification, reviewers might be tempted to think that your proposed research questions are not worth asking. Carefully justify the novelty of your idea. The more novel it is, the more experimental evidence you will need to support it.

Once you settle on the specific focus of your proposal, do not proceed until you answer the following questions to your full satisfaction.

- Does your idea have a clear focus? If one has to name a single factor that can kill a grant or fellowship proposal, it is lack of focus. Being overambitious is a common, and fatal, flaw of research proposals written by researchers in the early stages of their careers.
- Is your idea really that novel? Most important, does it duplicate similar projects currently funded by the same agency? Check the

agency's Web site and consult with their program administrators to answer this important question.

- Does the proposed research direction fit with the agenda of the funding agency?
- Are you sticking your neck out too far? Ideally, your idea should strike a balance between being novel and seeming safely doable. It is often difficult to convince reviewers to buy in to a risky and far-reaching research idea, even one that, if realized, might move the field forward by leaps and bounds. Some funding agencies, however, have specific budget allocations earmarked for such "high-risk, high-impact" proposals.
- In what way does your idea represent a better mousetrap than previous models? Could you convince others of your reasoning?
- Why hasn't someone else pursued this direction? Could it be that the idea is not worthwhile? Could answering the proposed question be beyond reach due to lack of necessary experimental means?
- Is your idea testable in a reasonable length of time? If you or others think you are proposing to do too much, seriously consider dividing the project into two proposals. This is a good situation to be in.
- Do you have training and expertise that would qualify you to perform the propose research project? Does your advisor?
- Is the proposed research a logical extension of your and your advisor's past or current research?
- Would answers to your questions have a significant impact on the specific field of knowledge?
- How passionate are you about the answering the questions posed in the proposal?
- Do your colleagues like your idea as well?

PRESENTING YOUR IDEAS CLEARLY

Many investigators become discouraged when they first realize that the quality of their presentation of a research idea has a significant impact on funding decisions. Isn't the research idea what really matters? The answer to this question is certainly yes, but most proposals are built on good ideas, so it is the quality of the presentations that enables reviewers to settle on a few to recommend for funding. Moreover, reviewers reasonably infer from clear and well-organized writing that the applicant thinks, and is likely to conduct research, in a clear and well-organized manner. Imagine yourself being offered the same car for the same price by two car dealerships that differ markedly not only in ambience, attitude, and salesmanship but in how clean and polished the cars are. I believe your choice would be obvious, given what these differences suggest about the quality of future mechanical service for your expensive investment.

The art of writing grant and fellowship proposals has, not surprisingly, been the topic of many textbooks and articles; a few good examples are listed at the end of this chapter. There also many helpful tools available online.

- National Institutes of Health, http://grants.nih.gov/grants/documentindex.html
- National Science Foundation, http://www.nsf.gov/funding/preparing/
- Foundation Center, http://foundationcenter.org/getstarted/learnabout/proposalwriting.html

Putting your ideas into words

Most of the following tips for writing successful grant proposals parallel the advice given in chapter 21 on general writing skills.

Needless to say, read the instructions, and read them carefully. Never rely on instructions or application forms you have previously copied and filed away. Funding agencies constantly revise and update their forms and instructions. Always get the latest version online.

Design the proposal using the "inverted pyramid" story structure typical of newspapers. Make your opening statement as informative as possible about the entire proposal, then elaborate. For example, the proposal's abstract should succinctly spell out the research problem and background, the hypothesis to be tested, and your plan for testing it. It should also state the impact of the proposed work. Provide clear and informative headings, diagrams, flowcharts, graphs, summaries and full narratives. These will allow reviewers to skim quickly through headings and diagrams to get an initial impression of what the proposed project is all about. This, you hope, will spark their interest to read the details. Flipping through these diagrams and summaries will also enable nonspecialist members of the panel to grasp the essence of the proposal. This approach makes it much more likely that you will succeed in delivering your message to all, albeit at different depths.

Aim for love at first sight. Design your writing to grab attention and induce enthusiasm as soon as the reviewer reads the first few paragraphs. This can be achieved by succinctly answering the major questions discussed earlier in this chapter: What? Why? How? Who? Where? and Who cares?

Demonstrate your enthusiasm and excitement about your proposed project, both overtly and covertly. Enthusiasm is often infectious.

Keep your overworked and mixed-expertise audience in mind, and keep the writing clear.

Use only conventional abbreviations, and not too many of those. Too many abbreviations can confuse reviewers, particularly those with little expertise in your specific research field. Providing a list of abbreviations is not sufficient, since it would require the reviewer to shuffle back and forth to interpret what you are talking about. The likely result is frustration and a sense of disconnectedness from the proposal.

Once you have the reviewers' attention, keep them hooked with smooth and logical flow and transitions. Do not force the reviewers to stop and think about how one part of the proposal logically relates to others. Do your best to make each section of your proposal a stand-alone mini-version of the entire text. When you present background knowledge, for example, briefly hint at the research questions you will ask in a later segment of the proposal. Recognize that reviewers might not read your entire proposal in one sitting, but rather in bits and pieces as their other obligations allow. This is often the situation, particularly in the case of lengthy proposals that run to tens of pages. By adopting this writing style, you will keep the reviewer reminded of what they read the day before and provide continuity with the part they are reading today. Doing this without appearing redundant requires grantsmanship skills that you'll develop over time.

Make sure you do not lose your focus. Constantly prune the text to remove extraneous parts that might distract the reviewer from the main goal of the proposal.

Consult with your peers and with colleagues who have more grant-writing experience. Ask them to read and honestly critique your drafts. Specifically ask them to identify any parts that they had to read more than once to fully understand. Go back to your keyboard and revise for additional clarity. Remember, reviewers likely will not have the time to read proposals more than once. More importantly, you will not be present at the review meeting to explain what you actually meant to say. By bearing these realities in mind you will avoid the temptation of feeling defensive when critiqued and not making needed revisions.

It takes significant time and commitment to achieve these goals. It is not uncommon for an applicant, even an experienced one, to spend a couple of months on a large grant proposal. Start early. Make the writing process more enjoyable and productive by writing small parts every day instead of very long writing sessions that are far apart. Writing an outstanding grant or fellowship proposal is very much like working on an oil painting. Getting away from the painting allows you to more clearly see what needs improvement. You should not expect to achieve perfection, or anywhere near it, when you write the first draft of each section of the proposal. Allow the flow of ideas in your mind to proceed uninterrupted. You will have ample opportunities for improvement and fine tuning.

Main components of a grant proposal

The instructions issued by various funding agencies may use different nomenclature and specify different sequences for the various sections of a grant proposal. In each case, however, you will be asked to provide a title for your proposal, a summary, a review of background knowledge, a statement of the specific aims and significance of the proposed project, a description of the methods you will use to address the research problem, and a detailed budget. Most will ask for a biographical sketch and a description of research facilities. Fellowship applications will also ask for your advisor's biographical sketch and history of training predoctoral students and postdoctoral fellows. You need not write these components of the proposal in the sequence they will follow in the final document. In fact, you must start with the heart of the application, that is, the specific questions and aims of the proposed project. This section often requires significant revisions and modifications during the course of writing, which in turn necessitates modification of other sections, particularly research design. Once this part is perfected, writing other components will be easy. The summary/abstract of the proposal is very influential in establishing a reviewer's attitude toward the proposal since it usually comes at the very beginning. However, one should write this section last since its purpose is to summarize the entire proposal.

FIRST TASK: SPECIFIC AIMS

You should start writing this part of the grant or fellowship proposal as early as possible, perhaps two months before the submission deadline. This section conveys to reviewers the research problem/question you propose to address. The best approach is to lay out the logic you followed in deciding on the specific set of research questions and evaluating their significance for the field. We suggest that you start with an introductory paragraph that addresses the following points.

- What are the main short- and long-range goals of the proposed project?
- What is already known?
- What critical information is missing?
- How vital are these gaps in knowledge?

This introductory paragraph becomes more significant in the case of granting agencies (e.g., the National Institutes of Health) that illogically instruct applicants to state their specific aims before providing the background that justifies the aims. Starting your proposal with this brief introduction circumvents this problem.

Next, list the questions you are proposing to address. The number and complexity of questions/specific aims naturally depends on the duration of the project and maximum budget allowable by the granting agency. You do not want to be accused of being unrealistic and overreaching, but you do not want to propose too little either. One specific aim or major question per year of work is usually reasonable. Individual aims should have the same focus and equal importance. Do not have one major aim and others that might appear to simply be add-ons.

End this section with a brief discussion of the expected gain in knowledge following completion of your studies. Address the following questions.

- Why do you think your idea is novel?
- Why are you especially qualified to address this specific research problem?
- In the case of a pre- or postdoctoral fellowship, why can your advisor be expected to provide you with outstanding research training?
- Does your institution provide a research environment that is supportive of your research training goals?
- What is the payoff when you complete your work? Specifically, how will your project benefit the mission of the particular funding agency you are applying to?

Note that this approach to the "specific aims" portion of your proposal addresses all the major questions we've discussed; once again: What? Why? How? Who? Where? Who cares?

Ask colleagues to critique this "master plan" and revise it accordingly. Do not proceed any further until you are fully convinced that this is the best plan of action. Consider this the sketch of an oil painting. It is much easier to modify the major objects or subjects the final painting will contain at this early stage.

SECOND TASK: BACKGROUND AND SIGNIFICANCE OF THE PROJECT

This section of the proposal is not meant to be a general or exhaustive review of the topic. Instead, it should provide a concise and logical summary of the background information that led you to formulate the proposed research questions. Make sure you have searched the existing literature carefully, paying particularly attention to recent publications in the field. To make this section easy to read, divide it into subsections with underlined headings. This will allow a selective reviewer to pick and choose which information to read. Assure a smooth flow between one subsection and the next, and do your best to refer to the link between this information and the research questions you are planning to pursue.

Maintain an unbiased and balanced approach in presenting points of view that either support or contradict your own. Do not appear unduly wedded to one camp while ignoring or minimizing others, unless you present strong evidence in support of your bias.

End this section with a succinct account of the significance of your proposal in terms of its potential to fill important gaps in knowledge. If possible, provide a simple table that highlights knowns, unknowns (knowledge gaps), and what you propose to add. As the saying goes, a picture is worth a thousand words.

Say, for example, that you propose to study the effects of exercise on production of nitric oxide (NO). The rationale is that both exercise and NO increase blood flow (BF) to muscles. You might summarize the major elements your project's rationale and goals in the following diagram.

What's known	Knowledge gap	Proposed question
a. Exercise increases BF b. NO increases BF	Link between exercise and NO production	Does exercise increase generation of NO?

THIRD TASK: PRELIMINARY DATA OR INITIAL WORK

This important section serves multiple purposes: it provides support for your hypothesis or research direction, demonstrates the feasibility of the proposed studies, and indicates your and your advisor's mastery of the methods, techniques, and approaches to be used in the proposed studies.

Some granting agencies specifically request a section presenting the results of your initial experiments or work related to the proposed project. Even when this is not the case, it behooves you to provide one; it will fit naturally within the section on background information. Avoid the use of appendixes to present your preliminary data. Appendixes may be ignored or even viewed with suspicion by reviewers who regard them as a ploy to get around imposed page limits.

Be brief, yet clear and informative. Divide this part into subsections with clear headings. Indicate how each finding supports the specific aims of the proposed studies.

At the end of this section, summarize all the preliminary findings and how they support your contemplated research direction and the feasibility of the proposed studies. Most important, clearly articulate why additional, more extensive studies are required. Do not expect the reviewers to reach this conclusion on their own. Be careful not to present so much preliminary data that a reviewer could conclude that you have already pretty much answered the proposed questions. There is a fine line between sufficient and excessive. This line will become more intuitively obvious as you gain more experience in grantsmanship.

Finally, provide a list of your (and/or your advisor's) publications that are specifically relevant to the proposal, even if a detailed biographical sketch is included elsewhere in the proposal. This adds credibility in your ability to conduct the proposed studies and the research training competence of your advisor.

Again, predoctoral fellowship applications require less preliminary data than postdoctoral ones. In both cases, however, excellence in research productivity and in the training history of your advisor are essential for the success of the proposal.

FOURTH TASK: DETAILS OF THE RESEARCH PLAN

By now, it will have been a while since the reviewers read about the specific aims and rationale of your project (particularly in case of a long application). You need to remind them and get them reconnected to the rationale of your proposal. Briefly restate your answers to the questions What? Why? and Who cares? before you go into details of How? Write a general, but concise, introductory paragraph about the specific and long-term goals of the project, the central hypothesis or question, supporting preliminary findings, and how your project will fill important knowledge gaps.

Then move on to the research plan. For each specific aim, state the hypothesis and summarize its rationale. Summarize preliminary data or findings that provide support for the question at hand. Provide details of the research approach and study design according to the following guidelines.

- Whenever possible, think of multiple complementary ways to address the specific aim or answer the question. Explain how the different strategies add together to provide a more comprehensive answer.
- Clearly explain the rationale for having chosen the approach you propose to follow over others. Be proactive in addressing questions that might come to a reviewer's mind regarding these choices.
- Think of all the controls necessary to validate the conclusions you hope to reach at the end of the proposed studies.
- Provide a summary of anticipated outcomes, if applicable, as suggested by your preliminary findings. State the importance of such findings to knowledge within the field. Better yet, lay out various possibilities and state how each might be interpreted. Mention how certain outcomes would influence the course of the studies and the final conclusions to be made. Do not appear wedded to a specific outcome unless there is sufficient support. This is another fine line to tread, since conclusive evidence in support of a specific outcome would obviate the need to perform the study.
- Think of potential obstacles and how you plan to circumvent them. Contemplate alternative approaches. This will highlight your careful and critical thinking. It also represents a prudent preemptive strike that should head off potential attacks by the reviewers. Be careful, though, not to shoot yourself in the foot by raising too much doubt about the feasibility of the proposed studies.

Repeat this sequence for each specific aim or question. At the end of this section, state how data will be analyzed (if applicable) and what criteria will be applied to interpret your findings. Also, provide a proposed timetable for each segment of the study. Careful justification of the time required will protect you against reviewers' temptation to reduce the duration of funding. The timetable will also give you a clearer sense of whether your plan to finish the proposed work in the allocated length of time is realistic. Immoderate ambition, common in proposals submitted by enthusiastic junior researchers, is a sure killer of grant proposals. One thing to bear in mind is that very few grant applications are criticized for proposing too little to do.

FIFTH TASK: THE FINALE

Now that you have grabbed the attention of reviewers and sustained their enthusiasm over so many pages of narrative, you want to end with a bang. As a reviewer reaches the end of your proposal, he or she likely has a pencil out and is ready to assign a score, hopefully a good one. Make that job easy by providing a succinct summary of your answers to the questions you've repeatedly addressed: What? Why? How? Who? Where? and So what? Very briefly list the research questions; describe your general approach to the research problem; explain why you and your advisor are uniquely positioned to perform the proposed studies; and state the significance and impact of the contemplated research on the field.

To assure a slam dunk, provide an answer to an important question that might occur to reviewers: Where will you go next? Summarize your future research plans, based on the possible outcomes of the proposed project. This demonstrates to the reviewers that your careful and critical thinking and planning will not stop at the end of the current project. You have the clear vision necessary to carry the ball even farther.

SIXTH TASK: THE SUMMARY OR ABSTRACT

We hope that by now you have realized the wisdom of waiting until the other parts of the proposal are in almost final form to write this introductory section. This ensures that your abstract will accurately reflect the finalized plan of action. Do your best in writing this section, since it is the one part of the proposal that all members of the review panel are likely to read. The guiding principles for writing the abstract are similar to those just described for writing the "finale."

Ask your colleagues for feedback

At this stage you hopefully still have a week or two until the submission deadline. Use this time wisely by asking a couple of colleagues to provide their honest critique of the proposal. Try to pick at least one person with expertise in your research field and another whose focus is in a related but distinctly different field. This will mimic the mixed levels of expertise of potential reviewers assigned to your proposal. Do not ask for this critique in the last minute; you will not get good feedback and will not have time to make major changes. Ask your colleagues to identify particularly any parts of the proposal that are difficult to follow or open to subjective interpretation. Ask them to point out segments that require better connection to the rest of the proposal. Take their criticisms and recommendations to heart, and revise your proposal accordingly.

Writing other components of the proposal

An opportune time to fill in these additional components is while you are waiting to hear back from your colleagues. While these final tasks are important, they do not require as much creativity as the main body of the proposal.

PROPOSAL TITLE

Decide on a short title that accurately describes the goals of your proposal. Avoid generalities and abbreviations, even if they are conventional in your particular field of study. Titles are often used by funding program administrators to assign review panels and specific reviewers to applications.

BIOGRAPHICAL SKETCH

Follow the funding agency's specific instructions for this section. Never attach your entire curriculum vitae. Reviewers read this part carefully to determine the expertise of an applicant in relation to the proposed research project. Design your biographical sketch accordingly, highlighting your most relevant publications if you have too many.

PROJECT BUDGET AND ITS JUSTIFICATION

This is usually a fun section to write, as it represents your wish list. Of course, in this case you have to justify everything you ask for. You should also pay careful attention to which specific categories of expenditure are permitted by the funding agency and which are not. Be realistic in assessing your needs, but also stay within reasonable limits. In particular, be cognizant of allowable budget ceilings and the average award size for this particular agency. Proposing an unreasonably high or low budget or number of years will betray your lack of experience in practical assessment of needs. Allowing some "fat" in the budget might be wise to minimize the detrimental consequences of likely budget reductions by reviewers or by the agency.

Getting local institutional approvals

Many first-time grant writers erroneously believe that a grant proposal is ready to go to the granting agency once they finish writing it. They do not realize that proposals are submitted on behalf of the institution rather than the individual investigator. This means that each proposal must be endorsed by the department chair, college dean, and the institutional office of research grants. In addition, departmental and institutional accountants are required to go through the proposal budget to ensure its compliance with regulations of the funding agency, to account for annual inflation, and so on. They also need to calculate the amount of indirect costs (overhead) to be allocated to the institution. This process naturally requires a few days or perhaps a week. Avoid last-minute surprises by showing parts of the proposal usually scrutinized by institutional officials—especially the budget—to your departmental accountant or grants manager a few weeks before the submission deadline. It also takes control out of the applicant's hands; these necessary final administrative steps are dependent on the availability of those who must sign off. One could miss the agency's deadline if something goes wrong and the final signatures are not obtained in a timely manner.

Another type of approval is necessitated if the research proposal requires the use of vertebrate animals or human subjects. Such study protocols at least have to be submitted to the appropriate institutional research compliance offices prior to sending the proposal to the granting agency. Actual approval of the appropriateness of these experimental protocols has to be obtained before the granting agency will release any research funds to the applicant.

Furthermore, certain types of grant proposals have to be accompanied by reference letters on behalf of the applicant. Ample time has to be allowed to avoid unduly putting references in a time crunch, and possibly missing the application deadline.

Final touches

Assure that your final product goes out in a perfect form. Do not let minor errors or glitches in appearance distract the reviewers' enthusiasm and attention. Remember, the smallest deduction from your score could make all the difference in the funding outcome.

Proofread the entire document carefully. Do not rely on your word processing program to do it for you. Such programs do not flag typographical errors if the end result is a correctly spelled, albeit unintended word; they will not recognize cases where "of" should read "if" or "know" should be "snow." Moreover, if the program you are using is set to automatically correct spelling, you run the risk of its choosing words completely different from the ones you had in mind. The errant word could deliver the wrong message, or at least generate distracting humor. It would be wise to turn these features off.

Finally, check the quality and completeness of content of all copies. Make sure there are no missing or smudged pages.

Revising unfunded proposals

Many grant or fellowship applications require at least one round of revision before they make it to the funding line. You should be as careful in responding to reviewers' critiques as you were when you wrote the original version of the proposal. Here are some guidelines to follow.

- Do not take your initial lack of success personally; do not get discouraged.
- Gather and include necessary new preliminary data.
- Provide a succinct introduction to highlight how you responded to reviewers' comments.
- Highlight major changes in the text (underlining, vertical lines in margin, etc.)
- Do not follow any of the reviewers' comments blindly. Stand up for you what you believe, firmly but gently. Your strong disagreement with specific comments by a reviewer may necessitate shredding a first draft you wrote in a moment of uncontrolled anger.
- Do not simply provide lip service; justify your agreement with the reviewers' suggestions. There is no guarantee that the same reviewer will read the application the second time around.
- Think twice before you change parts not criticized by the reviewers. This might open another can of worms.
- If the budget was reduced unjustifiably, defend your need for the full budget.
- Give your revision to colleagues to critique.

Take-home messages
- Write and submit proposals for predoctoral or postdoctoral fellowships even if you are fully supported by your program or advisor. This allows you to practice and, if funded, adds a feather in your hat.

- Identify all possible sources of funding in your field.
- Learn about the proposal review process and what reviewers look for.
- Fully understand the criteria used for proposal evaluation by your targeted funding agency.
- Ensure that your proposal is based on a novel questions and is clearly written and organized. These are the common feature of funded proposals.
- Aim to impress the reviewers in the first few pages, and to keep their attention until they read the last page.
- Design well-controlled experiments to answer your questions.
- Maintain a sharp focus. Relate every part of the proposal to the main question or hypothesis.
- Ask your advisor and colleagues for feedback prior to submission.

References and resources

Buscher, Leo F. 2005. "Everything You Wanted to Know about the NCI Grants Process but Were Afraid to Ask." U.S. Department of Health and Human Services, National Cancer Institute. http://www3.cancer .gov/admin/gab.

National Institutes of Health. "Computer Retrieval of Information on Scientific Projects." http://crisp.cit.nih.gov.

Foundation Center. "Proposal Writing Short Course." http://www.fdncenter.org/learn/shortcourse/prop1.html.

Kraicer, Jacob "Survival Skills for Graduate Students." http://www .physpharm.fmd.uwo.ca/undergrad/survivalwebv3/frame.htm.

Locke, Lawrence F., Waneen Wyrick Spirduso, and Stephen J. Silverman. 2007. *Proposals That Work: A Guide for Planning Dissertations and Grant Proposals.* 5th ed. Newbury Park, CA: Sage Publications.

National Institutes of Health Office of Extramural Research. "Resources for Grant Applicants." http://grants.nih.gov/grants/documentindex.html.

National Science Foundation. "How to prepare your proposal?" http:// www.nsf.gov/funding/preparing/.

Ogden, Thomas E., and Israel A. Goldberg. 2002. *Research Proposals: A Guide to Success.* 3rd ed. San Diego: Academic Press.

White, Virginia P. 1975. *Grants: How to Find Out about Them and What to Do Next.* New York: Plenum Press.

INDEX